Ben Stacy Jerrik (Ed.)

Galmpton, Torbay

AF307115

Ben Stacy Jerrik (Ed.)

Galmpton, Torbay

Torbay, Devon, Brixham, Greenway Estate, Listed building

Part Press

Imprint

Permission is granted to copy, distribute and/or modify this document under the terms of the GNU Free Documentation License, Version 1.2 or any later version published by the Free Software Foundation; with no Invariant Sections, with the Front-Cover Texts, and with the Back- Cover Texts. A copy of the license is included in the section entitled "GNU Free Documentation License".

All parts of this book are extracted from Wikipedia, the free encyclopedia (www.wikipedia.org).

You can get detailed informations about the authors of this collection of articles at the end of this book. The editors (Ed.) of this book are no authors. They have not modified or extended the original texts.

Pictures published in this book can be under different licences than the GNU Free Documentation License. You can get detailed informations about the authors and licences of pictures at the end of this book.

The content of this book was generated collaboratively by volunteers. Please be advised that nothing found here has necessarily been reviewed by people with the expertise required to provide you with complete, accurate or reliable information. Some information in this book maybe misleading or wrong. The Publisher does not guarantee the validity of the information found here. If you need specific advice (f.e. in fields of medical, legal, financial, or risk management questions) please contact a professional who is licensed or knowledgeable in that area.

Any brand names and product names mentioned in this book are subject to trademark, brand or patent protection and are trademarks or registered trademarks of their respective holders. The use of brand names, product names, common names, trade names, product descriptions etc. even without a particular marking in this works is in no way to be construed to mean that such names may be regarded as unrestricted in respect of trademark and brand protection legislation and could thus be used by anyone.

Cover image: www.ingimage.com
Concerning the licence of the cover image please contact ingimage.

Publisher:
Part Press is a trademark of
International Book Market Service Ltd., 17 Rue Meldrum, Beau Bassin, 1713-01 Mauritius
Email: info@bookmarketservice.com
Website: www.bookmarketservice.com

Published in 2012

Printed in: U.S.A., U.K., Germany. This book was not produced in Mauritius.

ISBN: 978-613-6-40088-4

Contents

Articles

References

Galmpton,_Torbay

Galmpton	
Galmpton	
Galmpton shown within Devon	
OS grid reference	SX888562
Unitary authority	Torbay
Ceremonial county	Devon
Region	South West
Country	England
Sovereign state	United Kingdom
Post town	BRIXHAM
Postcode district	TQ5
Dialling code	01803
Police	Devon and Cornwall
Fire	Devon and Somerset
Ambulance	South Western
EU Parliament	South West England
UK Parliament	Totnes

Galmpton is a village in Torbay, Devon, England. It is located west of Brixham in the historic civil parish of Churston Ferrers and is the home of Galmpton United F.C.. Greenway Estate, a Grade II* listed house and garden acquired by the National Trust in 1999, is nearby. The Paignton and Dartmouth Steam Railway has a station at Churston less than a mile from Galmpton.

Schools

- Churston Ferrers Grammar School
- Galmpton C of E Primary School

Torbay

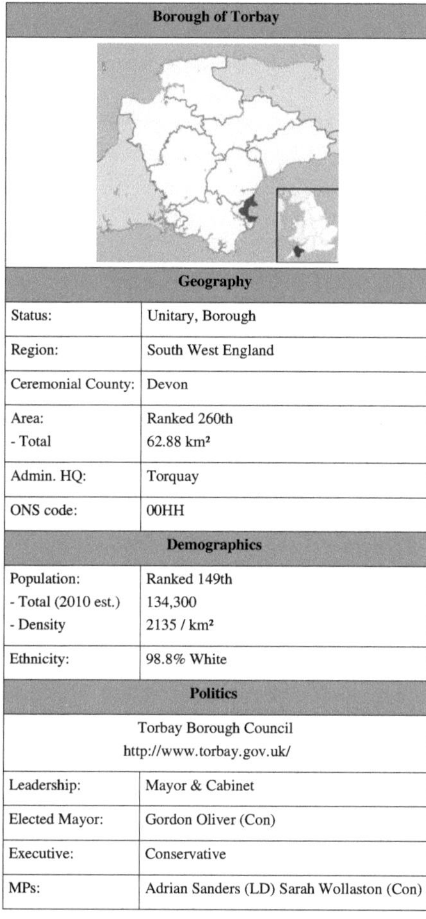

Borough of Torbay	
Geography	
Status:	Unitary, Borough
Region:	South West England
Ceremonial County:	Devon
Area: - Total	Ranked 260th 62.88 km²
Admin. HQ:	Torquay
ONS code:	00HH
Demographics	
Population: - Total (2010 est.) - Density	Ranked 149th 134,300 2135 / km²
Ethnicity:	98.8% White
Politics	
Torbay Borough Council http://www.torbay.gov.uk/	
Leadership:	Mayor & Cabinet
Elected Mayor:	Gordon Oliver (Con)
Executive:	Conservative
MPs:	Adrian Sanders (LD) Sarah Wollaston (Con)

Torbay (/tɔrˈbeɪ/) is an east-facing bay and natural harbour, at the westernmost end of Lyme Bay in the south-west of England, roughly equidistant from the cities of Exeter and Plymouth. Consisting of the towns of Torquay, Paignton and Brixham, and part of the ceremonial county of Devon, Torbay was made a unitary authority on 1 April 1998. Tourist authorities call it the "English Riviera" on account of its beaches and mild climate; it is a popular area with holiday makers.

History

That Torbay has been inhabited since paleolithic times is shown by human bones and tools found in Kents Cavern in Torquay. A maxilla fragment known as Kents Cavern 4 may be the oldest example of a modern human in Europe, dating back to 37,000–40,000 years ago.[1] [2] Roman soldiers are known to have visited Torquay during the period when Britain was a part of the Roman Empire, leaving offerings at a curious rock formation in Kent's Cavern, known as "The Face".

Both Brixham and Paignton appear in the Domesday Book of 1086 and Paignton was given the status of a borough having a market and fair in 1294.[3] The first major building in Torquay was Torre Abbey, a Premonstratensian monastery founded in 1196.[4]

William Prince of Orange (afterwards King William III) landed in Brixham on 5 November 1688, during the Glorious Revolution, and issued his famous declaration *"The Liberties of England and The Protestant Religion I Will Maintain"*.

Torquay's economy was, like Brixham's, initially based on fishing and agriculture, but in the early 19th century it began to develop into a fashionable seaside resort, initially frequented by members of the Royal Navy during the Napoleonic Wars while the Royal Navy anchored in Torbay and later, as the town's fame spread, by Victorian society.

The historic part of Paignton is inland: the low-lying coastal fringe was originally salt marsh. Kirkham House is a late medieval stone house which is open to the public at certain times of year,[5] and the Coverdale Tower adjacent to Paignton Parish Church is named after Bishop Miles Coverdale, who published an English translation of the Bible in 1536. Paignton remained a small fishing village until the early 19th century; a new harbour was built here in 1837.

The second phase in the expansion of Torbay began when Torre railway station was opened in December 1848. The railway was extended to Paignton in 1859 and to Brixham in 1861. As a result of its expansion, Torquay was granted borough status in 1872, and 1902 saw its first marketing campaign to summer tourists.

Torbay hosted the sailing events for the 1948 Summer Olympics in London.[6]

The County Borough of Torbay was created in 1968 by the amalgamation of the Municipal Borough of Torquay, Urban District of Paignton and Urban District of Brixham, also taking in parts of the civil parishes of Coffinswell and Kerswells from Newton Abbot Rural District and Churston Ferrers and Marldon from Totnes Rural District. The County Borough became the Borough of Torbay under local government reorganisation in 1974. It was made a unitary authority on 1 April 1998 making it responsible for its own affairs.

Governance

The area is represented nationally at the House of Commons by two MPs. Torquay (along with part of Paignton) is in the Torbay parliamentary constituency which was created in 1974 and has been held by Adrian Sanders of the Liberal Democrats since 1997. Brixham and part of Paignton fall within the Totnes constituency, with Conservative Sarah Wollaston elected. Torbay is in the South West England constituency of the European Parliament, together with the rest of South West England and Gibraltar.

Torbay Council is headed by the first directly elected mayor in the South West region. Conservative candidate Nicholas Bye became the first mayor elected under this system in October 2005, under an electoral system which was later described as "a total failure" with Bye receiving votes from fewer than 7% of the electorate.[7] He beat Liberal Democrat Nicholas Pannell in the second round of counting with a total of 7,096 votes to Pannell's 5,197. After the election, Bye noted the general apathy towards the concept displayed during the election, stating: "it is quite clear from canvassing that a lot of people did not want an elected mayor." The most recent mayoral election took place on 5 May 2011, in which Bye lost to Gordon Oliver.

For local elections the district is divided into 15 wards.[8] The Council elects 36 councillors in elections held every four years. Since the Torbay Council election, 2011, the council has a Conservative majority. The composition as of June 24 2011 was:

Party[9]	Seats
Conservative	21
Liberal Democrat	10[10]
Independent	3
Labour	1
UKIP	1

Geography

There are three main towns around the bay: Torquay in the north, Paignton in the centre, and Brixham in the south, which have become connected over the years, swallowing up villages and towns such as St Marychurch, Cockington, Marldon, Churston Ferrers and Galmpton. Torbay is bordered by the South Hams to the south and west, and by Teignbridge to the north. Nearby towns include Totnes and Dartmouth in the South Hams, and Newton Abbot and Teignmouth in Teignbridge.

Looking towards Paignton from Torquay. Torbay palms in the foreground.

The southern limit of Torbay is Berry Head, and the northern limit is Hopes Nose, although Torquay itself stretches further north into Babbacombe Bay, where the beaches at Oddicombe and Babbacombe can be found; these are noted for their interesting Breccia cliffs. Torbay's many geological features have led to the establishment of the English Riviera Geopark; as of July 2008, this is the sole urban geopark of the 53 geoparks worldwide.[11]

Because of the mild climate, Torbay palm trees are a common sight along the coast. However, this 'palm' is in fact a cabbage tree (*Cordyline australis*), originating from New Zealand. These trees flourish elsewhere in the UK. It is suggested that the popularity of cabbage trees in Torbay is attributable to their first being introduced to the UK in that region.

Economy

Torbay's main industry is tourism. It has a large number of European students learning English.

The fishing port of Brixham is home to one of England and Wales' most successful fishing fleets and regularly lands more value than any UK port outside Scotland. It is also a base for Her Majesty's Coastguard and the Torbay Lifeboat Station.

Torbay has been twinned with Hameln in Lower Saxony, Germany since 1973; and with Hellevoetsluis in the Netherlands since 1989.

Transport

Torbay lacks direct motorway links and is primarily served by the A380 road from Exeter. The last stretch of road, from Newton Abbot via Kingskerswell is mostly single carriageway and is often congested in summer and during commuter hours. There have been plans to remedy the situation though alternative routes were difficult to find because the road passes by areas of outstanding natural beauty.[12] A bypass had been planned since as early as the 1930s and has come close on a few occasions to being built: the latest plan was halted in autumn 2008,[13] and in October 2010 it was confirmed that the bypass was not one of the 24 schemes that had been approved.[14] However in November 2011 the government awarded £74.6 million towards the cost of the bypass, and the county council stated that it hoped that construction would start in October 2012 with completion in December 2015.[15]

Torbay's other main road links are the A379, which follows a coastal route from Teignmouth, passes through Torquay and Paignton, then goes on to Dartmouth; and the A385 which goes inland to Totnes and the A38.

The bus franchise is largely operated by Stagecoach South West of the similarly named group, that operates a large share of the market in Torbay and the neighbouring towns of South Devon. Open top buses are operated by Stagecoach and other companies in the summer.

An open top bus advertising the "English Riviera"

Torbay has three stations on the National Rail network, operated by First Great Western: Torre railway station is inland on the road from Torquay to Newton Abbot, Torquay railway station is close to Torre Abbey Sands and Paignton railway station serves that town and links with the heritage Paignton and Dartmouth Steam Railway to Kingswear connecting via the Dart ferry to Dartmouth.

Notable people

Famous former residents of Torbay include authors Agatha Christie (who set many of her novels in a thinly disguised version of the borough), Charles Kingsley, Edmund Gosse and Rudyard Kipling. Peter Cook, comic, (half of a famous comedy team with Dudley Moore); the great industrialist and architect of the nearby Atmospheric railway, Isambard Kingdom Brunel; Prog-rock band Wishbone Ash, supermodel Lily Cole, comedian Jim Davidson, tennis player/TV presenter Sue Barker, and astrologer Russell Grant also originate from the area.

Notes

[1] John R. Pike, *Torquay* (Torquay: Torbay Borough Council Printing Services, 1994), 5-6

[2] Rincon, Paul (2005-04-27). "Jawbone hints at earliest Britons" (http://news.bbc.co.uk/2/hi/science/nature/4482679.stm). news.bbc.co.uk. . Retrieved 2006-11-07.

[3] Parnell, Peggy (2007). *A Paignton Scrapbook*. Sutton Publishing. ISBN 978-0-7509-4739-8.

[4] Percy Russell, A History Of Torquay (Torquay: Devonshire Press Limited, 1960), p.19

[5] Kirkham House : Devon : South West : View properties : Properties : Days Out & Events : English Heritage (http://www.english-heritage. org.uk/server/show/conProperty.273)

[6] 1948 Summer Olympics official report. (http://www.la84foundation.org/6oic/OfficialReports/1948/OR1948.pdf) p. 50.

[7] "Mayor voting system is condemned" (http://news.bbc.co.uk/1/hi/england/devon/4374914.stm). *BBC News*. 2005-10-25. . Retrieved 2008-04-03.

[8] Torbay's wards are Berry Head-with-Furzeham (3 councillors), Blatchcombe (3 councillors), Churston Ferrers-with-Galmpton (2 councillors), Clifton-with-Maidenway (2 councillors), Cockington-with-Chelston (3 councillors), Ellacombe (2 councillors), Goodrington-with-Roselands (2 councillors), Preston (3 councillors), Roundham-with-Hyde (2 councillors), St Marychurch (3 councillors), St. Mary's-with-Summercombe (2 councillors), Shiphay-with-The Willows (2 councillors), Tormohun (3 councillors), Watcombe (2 councillors), and Wellswood (2 councillors)

[9] "England council elections" (http://www.bbc.co.uk/news/special/election2011/council/html/hh.stm). *BBC News*. .

[10] "Declaration of result of poll" (http://www.torbay.gov.uk/index/council/elections/electionresults/byelectionresults.htm). *Torbay Council*. . Retrieved 2011-06-24.

[11] Global status for Torbay (http://www.bbc.co.uk/devon/content/articles/2007/09/17/torbay_geopark_status_feature.shtml) (retried 7 July 2008)

[12] "Impact of the First Local Travel Plan on Torbay" (http://web.archive.org/web/20070928024909/http://www.torbay.gov.uk/ltp1_delivery_report_final_document.pdf). torbay.gov.uk. Archived from the original (http://www.torbay.gov.uk/ltp1_delivery_report_final_document.pdf) on 2007-09-28. . Retrieved 2006-12-02.

[13] Herald Express 26th November 2008

[14] "Kingskerswell bypass 'not on government transport list'" (http://www.bbc.co.uk/news/uk-england-devon-11629560). BBC News - Devon. 2010-10-26. . Retrieved 2010-12-14.

[15] "Kingskerswell bypass work could start in 2012" (http://www.bbc.co.uk/news/uk-england-devon-15951459). BBC News. 2011-11-29. . Retrieved 2011-11-30.

References

External links

- Torbay Council (http://www.torbay.gov.uk)
- The Official Tourist Board (http://www.englishriviera.co.uk)
- Torbay (http://www.dmoz.org//Regional/Europe/United_Kingdom/England/Devon/Torbay//) at the Open Directory Project

Devon

<table>
<tr><td colspan="2" align="center">Devon</td></tr>
<tr><td colspan="2" align="center">Flag</td></tr>
<tr><td colspan="2" align="center">Motto of County Council: Auxilio divino (Latin: By divine aid)</td></tr>
<tr><td colspan="2" align="center">Geography</td></tr>
<tr><td>Status</td><td>Ceremonial & (smaller) Non-metropolitan county</td></tr>
<tr><td>Region</td><td>South West England</td></tr>
<tr><td>Area
- Total
- Admin. council
- Admin. area</td><td>Ranked 4th
6707 km^2 (unknown operator: u'strong' sq mi)
Ranked 3rd
6564 km^2 (unknown operator: u'strong' sq mi)</td></tr>
<tr><td>Admin HQ</td><td>Exeter</td></tr>
<tr><td>ISO 3166-2</td><td>GB-DEV</td></tr>
<tr><td>ONS code</td><td>18</td></tr>
<tr><td>NUTS 3</td><td>UKK43</td></tr>
<tr><td colspan="2" align="center">Demography</td></tr>
<tr><td>Population
- Total (2010 est.)
- Density
- Admin. council
- Admin. pop.</td><td>Ranked 11th
1,143,000
170 /km^2 (unknown operator: u'strong' /sq mi)
Ranked 12th
750,000</td></tr>
<tr><td>Ethnicity</td><td>98.7% White British</td></tr>
<tr><td colspan="2" align="center">Politics</td></tr>
<tr><td colspan="2" align="center">Devon County Council
http://www.devon.gov.uk</td></tr>
<tr><td>Executive</td><td>Conservative</td></tr>
</table>

Members of Parliament	• Ben Bradshaw (L)
	• Oliver Colvile (C)
	• Geoffrey Cox (C)
	• Nick Harvey (LD)
	• Anne-Marie Morris (C)
	• Neil Parish (C)
	• Adrian Sanders (LD)
	• Alison Seabeck (L)
	• Gary Streeter (C)
	• Mel Stride (C)
	• Hugo Swire (C)
	• Sarah Wollaston (C)

Districts

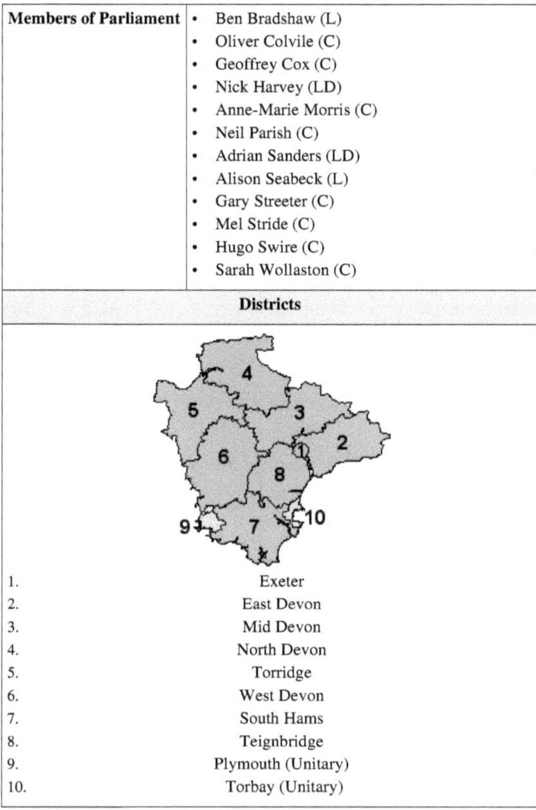

1.	Exeter
2.	East Devon
3.	Mid Devon
4.	North Devon
5.	Torridge
6.	West Devon
7.	South Hams
8.	Teignbridge
9.	Plymouth (Unitary)
10.	Torbay (Unitary)

Devon (◀ /ˈdɛvən/) is a large county in southwestern England. The county is sometimes referred to as **Devonshire**, although it has never been officially "shired" and that use often indicates a traditional or historical context. Nowadays, "Devonshire" is mostly a demonym.

The county shares borders with Cornwall to the west and Dorset and Somerset to the east. Its southern coast abuts the English Channel and its northern coast the Bristol Channel and Celtic Sea. The name "Devon" derives from the ancient Dumnonia, which was home to the independent kingdom of Brythonic Celtic speaking people who inhabited this area of the southwestern peninsula of Britain continuously from through the Roman era until partial absorption into the English-speaking Kingdom of Wessex some time in the eighth or ninth century, with some emigration to the sister Kingdom of Domnonee in Brittany taking place.

Devon is the fourth largest of the English counties by area and has a population of 1,141,600 making it the 11th largest. The county town is the cathedral city of Exeter. In addition to Devon County Council, the county contains two unitary authorities (independent from Devon County Council's control): the port city of Plymouth and Torbay, a conurbation of seaside resorts. Plymouth is also the largest city in Devon. Much of the county is rural (including national park) land, with a low population density by British standards. It contains Dartmoor 954 km^2 (**unknown operator: u'strong'** sq mi), the largest open space in southern England.[1] It is the only English county to have two separate coastlines – a north and southern coastline.

The county is the location of part of England's only natural UNESCO World Heritage Site, the Dorset and East Devon Coast, known as the Jurassic Coast for its geology and geographical features. It is also home to Braunton

Burrows UNESCO Biosphere Reserve, a dune complex in the north of the county. Along with Cornwall, Devon is known as the "Cornubian massif". This geology gives rise to the landscapes of Dartmoor and Exmoor, the latter two being national parks. Devon has seaside resorts and historic towns and cities, rural scenery and a mild climate, accounting for the large tourist sector of its economy.

History

Toponymy

The name *Devon* derives from the name of the Celtic people who inhabited the southwestern peninsula of Britain at the time of the Roman invasion c. AD 50, known as the Dumnonii, thought to mean "deep valley dwellers" from proto Celtic **dubnos* 'deep'. In the Brythonic Celtic languages, Devon is known as *Dyfnaint* (Welsh), *Devnent* (Breton) and *Dewnens* (Cornish) each meaning 'deep valleys'. (For an account of Celtic Dumnonia see the separate article).

William Camden, in his 1607 edition of *Britannia*, described Devon as being one part of an older, wider country that once included Cornwall:

> THAT region which, according to the Geographers, is the first of all Britaine, and, growing straiter still and narrower, shooteth out farthest into the West, [...] was in antient time inhabited by those Britans whom Solinus called Dunmonii, [...] But the Country of this nation is at this day divided into two parts, knowen by later names of Cornwall and Denshire, [...]
>
> —William Camden, *Britannia*.[2]

The term "Devon" is normally used for everyday purposes e.g. "Devon County Council" but "Devonshire" continues to be used in the names of the "Devonshire and Dorset Regiment" and "The Devonshire Association". One erroneous theory is that the "shire" suffix is due to a mistake in the making of the original letters patent for the Duke of Devonshire, resident in Derbyshire. However, there are references to "Defenascire" in Anglo-Saxon texts from before 1000 AD (this would mean "Shire of the Devonians"),[3] which translates to modern English as "Devonshire". The term Devonshire may have originated around the 8th century, when it changed from Dumnonia (Latin) to Defenascir.[4]

Human occupation

Kents Cavern in Torquay had produced human remains from 30–40,000 years ago. Dartmoor is thought to have been occupied by Mesolithic hunter-gatherer peoples from about 6000 BC. The Romans held the area under military occupation for around 350 years. Later, the area began to experience Saxon incursions from the east around 600 AD, firstly as small bands of settlers along the coasts of Lyme Bay and southern estuaries and later as more organised bands pushing in from the east. Devon became a frontier between Brythonic Dumnonia and Anglo-Saxon Wessex, and it was largely absorbed into Wessex by the mid 9th century. The border with Cornwall was set by King Athelstan on the east bank of the River Tamar in 936 AD. Danish raids also occurred sporadically along many coastal parts of Devon between around 800AD and just before the time of the Norman conquest, including at Taintona (a settlement on the Teign estuary) in 1001.

Devon has also featured in most of the civil conflicts in England since the Norman Conquest, including the Wars of the Roses, Perkin Warbeck's rising in 1497, the Prayer Book Rebellion of 1549, and the English Civil War. The arrival of William of Orange to launch the Glorious Revolution of 1688 took place at Brixham.

Devon has produced tin, copper and other metals from ancient times. Devon's tin miners enjoyed a substantial degree of independence through Devon's Stannary Parliament, which dates back to the 12th century. The last recorded sitting was in 1748.[5]

Economy and industry

Like neighbouring Cornwall to the west, historically Devon has been disadvantaged economically compared to other parts of Southern England, owing to the decline of a number of core industries, notably fishing, mining and farming. Agriculture has been an important industry in Devon since the 19th century.[6] The 2001 UK foot and mouth crisis harmed the farming community severely.[7] Since then some parts of the agricultural industry have begun to diversify and recover, with a strong local food sector and many artisan producers. It is also the headquarters of the UK's largest organic veg box provider - Riverford Organics, based near Totnes and River Cottage based near Axminster. Nonetheless the dairy industry is still suffering from the low prices offered for wholesale milk by major dairies and especially large supermarket chains.

Part of the seafront of Torquay, south Devon, at high tide

The attractive lifestyle of the area is drawing in new industries which are not heavily dependent upon geographical location; Dartmoor, for instance, has recently seen a significant rise in the percentage of its inhabitants involved in the financial services sector. In 2003, the Met Office, the UK's national and international weather service, moved to Exeter and employs around 1,000 people. .

Since the rise of seaside resorts with the arrival of the railways in the 19th century, Devon's economy has been heavily reliant on tourism. The county's economy followed the declining trend of British seaside resorts since the mid-20th century, but with some recent revival and regeneration of its resorts, particularly focused around camping; sports such as surfing, cycling, sailing and heritage. This revival has been aided by the designation of much of Devon's countryside and coastline as the Dartmoor and Exmoor national parks, and the Jurassic Coast and Cornwall and West Devon Mining Landscape World Heritage Sites. In 2004 the county's tourist revenue was £1.2 billion..[8] More successful visitor attractions in recent years have tended to target higher spending tourists, particularly focusing around the food and drink sector and watersports. Examples such as the Venus Cafes; Damien Hirst's restaurant in Ilfracombe; Burgh Island; surfing and camping around Croyde and Woolacombe and sailing around Salcombe have all proved a big draw in recent years.

Incomes vary significantly across the county, with parts of Torridge and Torbay having among the lowest earnings in the whole of the UK. Incomes in much of the South Hams and in villages surrounding Exeter and Plymouth are by contrast above the national average. Likewise levels of deprivation tend to be highest in urban areas such as Torbay, Plymouth, plus parts of Exeter and Ilfracombe. They are lowest in the rural fringes of big settlements, easily commutable and leafy, such as Ivybridge, Woodbury, Kenton and Braunton.

Geography and geology

A sharp geological dividing line cuts across Devon roughly to the west of Tiverton and Exeter and ending around Newton Abbot and Torquay. This forms part of the Tees/Exe line dividing Britain from the lowlands (sedimentary rocks) that are predominant to the east of the line and the higher land (igneous and metamorphic rocks) which dominates to the west.

Heathland at Woodbury Common in south east Devon

The principal geological formations of Devon are the Devonian (in north Devon, south west Devon and extending into Cornwall); the Culm Measures (north western Devon also extending into north Cornwall); and the granite intrusion of Dartmoor in central Devon, part of the Cornubian batholith. There are small remains of pre-Devonian rocks on the south Devon coast.[9]

Devon gave its name to a geological period: the Devonian period, so named because of the abundance of the grey limestone found there. It was Roderick Murchison and Adam Sedgwick who originally named the Devonian Period following research they carried out in Devon, and in particular, Torbay. They found some unusual marine fossils in the limestone at Lummaton Quarry and it was this discovery that led to the time period becoming known globally as the Devonian.

Devon's second major rock system[10] is the Culm Measures, a geological formation of the Carboniferous period that occurs principally in Devon and Cornwall. The measures are so called either from the occasional presence of a soft, sooty coal, which is known in Devon as *culm*, or from the contortions commonly found in the beds.[11] This formation stretches from Bideford to Bude in Cornwall, and contributes to a gentler, greener, more rounded landscape. It is also found on the western, north and eastern borders of Dartmoor.

The whole of central Devon is occupied by the largest area of igneous rock in South West England, Dartmoor.

Ilfracombe, on the coast of North Devon.

The sedimentary rocks in more eastern parts of the county include Permian and Triassic sandstones (giving rise to Devon's well known fertile red soils); Bunter pebble beds around Budleigh Salterton and Woodbury Common and Jurassic rocks in the easternmost parts of Devon. Smaller deposits of even newer rocks also exist, such as Cretaceous chalk cliffs at Beer head and gravels on Haldon, plus Eocene and Oligocene ball clay and lignite deposits in the Bovey Basin, formed around 50m years ago under tropical forest conditions.

Devon is the only county in England to have two separate coastlines; the South West Coast Path runs along the entire length of both, around 65% of which is named as Heritage Coast.[12] Devon has more mileage of road than any other county in England: before the changes to counties in 1974 it was the largest by area of the counties not divided into two or three parts. (Its acreage was until 1974 1,658,288: only exceeded by the West Riding of Yorkshire.)[13] The islands of Lundy and Eddystone are also in Devon.

Inland, the Dartmoor National Park lies wholly in Devon, and the Exmoor National Park lies in both Devon and Somerset. Apart from these areas of high moorland the county has attractive rolling rural scenery and villages with thatched cob cottages. All these features make Devon a popular holiday destination.

In South Devon the landscape consists of rolling hills dotted with small towns, such as Dartmouth, Ivybridge, Kingsbridge, Salcombe, and Totnes. The towns of Torquay and Paignton are the principal seaside resorts on the south coast. East Devon has the first seaside resort to be developed in the county, Exmouth and the more upmarket Georgian town of Sidmouth, headquarters of the East Devon District Council. Exmouth marks the western end of the Jurassic Coast World Heritage Site. Another notable feature is the coastal railway line between Newton Abbot and the Exe Estuary: the red sandstone cliffs and sea views are very dramatic and in the resorts railway line and beaches are very near.

North Devon is very rural with few major towns except Barnstaple, Great Torrington, Bideford and Ilfracombe. Devon's Exmoor coast has the highest cliffs in southern Britain, culminating in the Great Hangman, a 318 m (**unknown operator: u'strong'** ft) "hog's-back" hill with an 250 m (**unknown operator: u'strong'** ft) cliff-face, located near Combe Martin Bay.[14] Its sister cliff is the 218 m (**unknown operator: u'strong'** ft) Little Hangman, which marks the western edge of coastal Exmoor. One of the features of the North Devon coast is that Bideford Bay and the Hartland Point peninsula are both west-facing, Atlantic facing coastlines; so that a combination of an off-shore (east) wind and an Atlantic swell produce excellent surfing conditions. The beaches of Bideford Bay (Woolacombe, Saunton, Westward Ho! and Croyde), along with parts of North Cornwall and South Wales, are the main centres of surfing in Britain.

Climate

Devon generally has a mild climate, heavily influenced by the North Atlantic Drift. In winter snow is relatively uncommon away from high land, although there are exceptions, such as the snowfalls of February 2009, and December 2010. The county has warm summers with occasional hot spells and cool rainy periods. Winters are generally mild and the county often experiences some of the mildest winters in the world for its latitude, along with Brittany and Vancouver, Canada, with average daily maximum temperatures in January approaching 10 degrees C. Rainfall varies significantly across the county, ranging from over 2000 mm (**unknown operator: u'strong'** in) on parts of Dartmoor, to around 750 mm (**unknown operator: u'strong'** in) in the rain shadow along the coast in southeastern Devon and around Exeter. Sunshine amounts also vary widely: the moors are generally cloudy, but the SE coast from Salcombe to Exmouth is one of the sunniest parts of the UK. In summer, easterly or southeasterly winds mean the area around Saunton and Croyde often records among the highest temperatures in Britain. Similarly, with westerly or southwesterly winds and high pressure the area around Torbay and Teignmouth will often be very warm, with long sunny spells due to shelter by high ground (Foehn wind).

Fields in south Devon after a snowfall.

Ecology

The variety of habitats means that there is a wide range of wildlife (see Dartmoor wildlife, for example). A popular challenge among birders is to find over 100 species in the county in a day. The county's wildlife is protected by several wildlife charities such as the Devon Wildlife Trust, a charity which looks after 40 nature reserves. The Devon Bird Watching and Preservation Society (DBWPS) is a county bird society with a long and distinguished history dating back to 1928. It is dedicated to the study and conservation of wild birds looks after several areas, such as Beesands Ley. There is also the RSPB, which has reserves in the county, as well as English Nature, who look after several reserves such as Dawlish Warren. The botany of the county is

Ponies grazing on Exmoor near Brendon, North Devon

very diverse and includes some rare species not found elsewhere in the British Isles other than Cornwall. Botanical reports begin in the 17th century and there is a *Flora Devoniensis* by Jones and Kingston in 1829, and a *Flora of Devon* in 1939 by Keble Martin and Fraser.[15] [16] There is a general account by W. P. Hiern and others in *The Victoria History of the County of Devon*, vol. 1 (1906); pp. 55–130, with map. Devon is divided into two Watsonian vice-counties: north and south, the boundary being an irregular line approximately across the higher part of Dartmoor and then along the canal eastwards.

Rising temperatures have led to Devon becoming the first place in modern Britain to cultivate olives commercially.[17]

Politics and administration

The administrative centre of Devon is the city of Exeter. The largest city in Devon, Plymouth, and the conurbation of Torbay (including Torquay, Paignton and Brixham) have been unitary authorities since 1998, separate from the remainder of Devon which is administered by Devon County Council for the purposes of local government.

Devon County Council is controlled by the Conservatives, and the political representation of its 62 councillors are: 41 Conservatives, 14 Liberal Democrats, four Labour, two Independents and one Green.[18] At a national level, Devon has seven Conservative MPs, two Liberal Democrat MPs, and two Labour MPs.

Exeter Cathedral

In December 2007, the Department for Communities and Local Government referred Exeter City Council's bid to become a Unitary Council to the Boundary Committee for England, as they felt the application did not meet all their strict criteria. The Boundary Committee was asked to look at the feasibility of a unitary Exeter in the context of examining options for unitary arrangements in the wider Devon county area, and reported back in July 2008 recommending a "unitary Devon" (excluding Plymouth and Torbay), with a second option of a "unitary Exeter & Exmouth" (combined) and a unitary "rest of Devon". These proposals were put out to consultation until September 2008 and the Committee was expected to make final recommendations to the Secretary of State by the end of the year. As a result of a number of legal challenges to the process and also dissatisfaction on the part of the Secretary of State with the manner in which the Boundary Committee is assessing proposals, it now looks likely that a recommendation will not be forthcoming until March or April 2009.[19]

Hundreds

Historically Devon was divided into 32 hundreds:[20] Axminster, Bampton, Black Torrington, Braunton, Cliston, Coleridge, Colyton, Crediton, East Budleigh, Ermington, Exminster, Fremington, Halberton, Hartland, Hayridge, Haytor, Hemyock, Lifton, North Tawton and Winkleigh, Ottery, Plympton, Roborough, Shebbear, Shirwell, South Molton, Stanborough, Tavistock, Teignbridge, Tiverton, West Budleigh, Witheridge, and Wonford.

Cities, towns and villages

The inner harbour, Brixham, south Devon, at low tide

The main settlements in Devon are the cities of Plymouth, a historic port now administratively independent, Exeter, the county town, and Torbay, the county's tourist centre. Devon's coast is lined with tourist resorts, many of which grew rapidly with the arrival of the railways in the 19th century. Examples include Dawlish, Exmouth and Sidmouth on the south coast, and Ilfracombe and Lynmouth on the north. The Torbay conurbation of Torquay, Paignton and Brixham on the south coast is now administratively independent of the county. Rural market towns in the county include Barnstaple, Bideford, Honiton, Newton Abbot, Okehampton, Tavistock, Totnes and Tiverton.

The boundary with Cornwall has not always been on the River Tamar as at present: until the late 19th century a few parishes in the Torpoint area were in Devon and five parishes now in north-east Cornwall were in Devon until 1974. (However for ecclesiastical purposes these were nevertheless in the Archdeaconry of Cornwall and in 1876 became part of the Diocese of Truro.)

Religion

Ancient and medieval history

Celtic and Roman practices were the first known religions in Devon, although in the first centuries AD, Christianity was introduced to Devon. In the Sub-Roman period the church in the British Isles was characterised by some differences in practice from the Latin Christianity of the continent of Europe and is known as Celtic Christianity;[21] however it was always in communion with the wider Roman Catholic Church. Many Cornish saints are commemorated also in Devon in legends, churches and placenames. Western Christianity came to Devon when it was over a long period incorporated into the kingdom of Wessex and the jurisdiction of the bishop of Wessex. Saint Petroc is said to have passed through Devon, where ancient dedications to him are even more numerous than in Cornwall: a probable seventeen (plus Timberscombe just over the border in Somerset), compared to Cornwall's five. The position of churches bearing his name, including one within the old Roman walls of Exeter (Karesk), are nearly always near the coast, reminding us that in those days travelling was done mainly by sea. The Devonian villages of Petrockstowe and Newton St Petroc are also named after Saint Petroc and the flag of Devon is dedicated to him.

The history of Christianity in the South West of England remains to some degree obscure. Parts of the historic county of Devon formed part of the diocese of Wessex, while nothing is known of the church organisation of the Celtic areas. About 703 Devon and Cornwall were included in the separate diocese of Sherborne and in 900 this was again divided into two, the Devon bishop having from 905 his seat at Tawton (now Bishop's Tawton) and from 912 at Crediton, birthplace of St Boniface. Lyfing became Bishop of Crediton in 1027 and shortly afterwards became Bishop of Cornwall.

The two dioceses of Crediton and Cornwall, covering Devon and Cornwall, were permanently united under Edward the Confessor by Lyfing's successor Bishop Leofric, hitherto Bishop of Crediton, who became first Bishop of Exeter under Edward the Confessor, which was established as his cathedral city in 1050. At first, the abbey church of St Mary and St Peter, founded by Athelstan in 932 and rebuilt in 1019, served as the cathedral.

Later history

In 1549, the Prayer Book Rebellion caused the deaths of thousands of people from Devon and Cornwall. During the English Reformation, churches in Devon officially became affiliated with the Church of England. The Methodism of John Wesley proved to be very popular with the working classes in Devon in the 19th century. Methodist chapels became important social centres, with male voice choirs and other church-affiliated groups playing a central role in the social lives of working class Devonians. Methodism still plays a large part in the religious life of Devon today, although the county has shared in the post-World War II decline in British religious feeling.

The Diocese of Exeter remains the Anglican diocese including the whole of Devon. The Roman Catholic Diocese of Plymouth was established in the mid 19th century.[22]

Judaism

Despite its small Jewish population, Devon is also noted for containing two of Britain's oldest synagogues, located in Plymouth and Exeter, built in 1762 and 1763 respectively.

Symbols

Coat of arms

There was no established coat of arms for the county until 1926: the arms of the City of Exeter were often used to represent Devon, for instance in the badge of the Devonshire Regiment. During the forming of a county council by the Local Government Act 1888 adoption of a common seal was required. The seal contained three shields depicting the arms of Exeter along with those of the first chairman and vice-chairman of the council (Lord Clinton and the Earl of Morley).[23]

On 11 October 1926, the county council received a grant of arms from the College of Arms. The main part of the shield displays a red crowned lion on a silver field, the arms of Richard Plantagenet, Earl of Cornwall. The *chief* or upper portion of the shield depicts an ancient

The coat of arms of Devon County Council

ship on wavers, for Devon's seafaring traditions. The Latin motto adopted was *Auxilio Divino* (by Divine aid), that of Sir Francis Drake. The 1926 grant was of arms alone. On 6 March 1962 a further grant of crest and supporters was obtained. The crest is the head of a Dartmoor Pony rising from a "Naval Crown". This distinctive form of crown is formed from the sails and sterns of ships, and is associated with the Royal Navy. The supporters are a Devon bull and a sea lion.[24] [25]

The County Council adopted a "ship silhouette" logo after the 1974 reorganisation, adapted from the ship emblem on the coat of arms, but following the loss in 1998 of Plymouth and Torbay re-adopted the coat of arms. In April 2006 the council unveiled a new logo which was to be used in most everyday applications, though the coat of arms will continue to be used for "various civic purposes".[26] [27]

Flag

Devon also has its own flag which has been dedicated to Saint Petroc, a local saint with dedications throughout Devon and neighbouring counties. The flag was adopted in 2003 after a competition run by BBC Radio Devon.[28] The winning design was created by website contributor Ryan Sealey, and won 49% of the votes cast. The colours of the flag are those popularly identified with Devon, for example, the colours of Exeter University, the rugby union team, and the Green and White flag flown by the first Viscount Exmouth at the Bombardment of Algiers (now on view at the Teign Valley Museum), as well as one of the county's football teams, Plymouth Argyle. On 17 October 2006, the flag was hoisted for the first time outside County Hall in Exeter to mark Local Democracy Week, receiving official recognition from the county council.[29]

Place names and customs

Devon's place names include many with the endings "coombe/combe" and "tor" — Coombe being the Brythonic word for "valley" or hollow (cf Welsh 'cwm') whilst tor derives from a number of Celtic loan-words in English (Old Welsh twrr and Scots Gaelic tòrr) and is used as a name for the formations of rocks found on the moorlands. Its frequency is greatest in Devon, where it is the second most common place name component (after 'ton', derived from the Old English *tun* meaning enclosure, farmstead, farm or village).

The beach at Westward Ho!, North Devon, looking north towards the Taw and the Torridge estuaries

Devon has a variety of festivals and traditional practices, including the traditional orchard-visiting Wassail in Whimple every 17 January and the carrying of flaming tar barrels in Ottery St. Mary, where people who have lived in Ottery for long enough are called upon to celebrate Bonfire Night by running through the village (and the gathered crowds) with flaming barrels of tar on their backs.[30] Berry Pomeroy still celebrates "Queen's Day" for Elizabeth I.

Education

Devon has a mostly comprehensive education system. There are 37 state and 23 independent secondary schools. There are three tertiary (FE) colleges and an agricultural college (Bicton College, near Budleigh Salterton). Torbay has 8 state (with 3 grammar schools) and 3 independent secondary schools, and Plymouth has 17 state (with 3 grammar schools — two female and one male) and one independent school, Plymouth College. East Devon and Teignbridge have the largest school populations, with West Devon the smallest (with only two schools). Only one school in Exeter, Mid Devon, Torridge and North Devon have a sixth form — the schools in other districts mostly have sixth forms, with all schools in West Devon and East Devon having a sixth form. The county also plays host to two major British universities, the University of Exeter (split between the Streatham Campus and St Luke's Campus both in Exeter and a campus in Cornwall); in Plymouth the University of Plymouth, the fourth largest university in Britain is present, along with the Marjon's College to the city's north. Both the universities of Exeter and Plymouth have co-formed the Peninsula College of Medicine and Dentistry which has bases in Exeter and Plymouth. There is also Schumacher College.

Cuisine

The county has given its name to a number of culinary specialities. The Devonshire cream tea, involving scones, jam and clotted cream, is thought to have originated in Devon (though claims have also been made for neighbouring counties); in other countries, such as Australia and New Zealand, it is known as a "Devonshire tea".[31] [32] [33] In Australia, Devon is a name for luncheon meat (processed ham).

In October 2008, Devon was awarded Fairtrade County status by the Fairtrade Foundation.

Sport

Devon has been home to a number of customs, such as its own form of wrestling. As recently as the 19th century, a crowd of 17,000 at Devonport, near Plymouth, attended a match between the champions of Devon and Cornwall. Another Devon sport was outhurling which was played in some regions until the 20th century (e.g. 1922, at Great Torrington). Other ancient customs which survive include Dartmoor step dancing, and "Crying The Neck".

Devon has three professional football teams, based in each of its three most populous towns and cities. In the 2011/2012 football season, Exeter City F.C. will compete in Football League One, and Torquay United F.C. and Plymouth Argyle F.C. in Football League Two. Plymouth's highest Football League finish was fourth in the Second Division which was achieved twice in 1932 and 1953 respectively, while Torquay and Exeter have never progressed beyond the third tier of the league. The county's biggest non-league club is Tiverton Town F.C. which competes in the Southern Football League Division One South & West.

Rugby Union is popular in Devon with two teams – Exeter Chiefs play in the Aviva Premiership and Plymouth Albion who are, as of 2011, in the RFU Championship. In basketball, Plymouth Raiders play in the British Basketball League. Tamar Valley Cannons, also based in Plymouth, are Devon's only other representatives in the National Leagues. Motorcycle speedway is also supported in the county, with both the Exeter Falcons and Plymouth Devils succeeding in the National Leagues in recent years.

There are four rugby league teams in Devon. Plymouth Titans, Exeter Centurions, Devon Sharks from Torquay and East Devon Eagles from Exmouth. They all play in the Rugby League Conference.

Devon also boasts a field hockey club who play in the National Premier League, the University of Exeter Hockey Club

Horse Racing, particularly point to point racing and National Hunt Racing is also popular in the county, with two National Hunt racecourses (Exeter and Newton Abbot), and numerous point to point courses. There are also many successful professional racehorse trainers based in Devon.

The county is represented in cricket by Devon County Cricket Club, who play at a Minor counties level.

Devonians

Devon is known for its mariners, such as Sir Francis Drake, Sir Humphrey Gilbert, Sir Richard Grenville, Sir Walter Raleigh, and Sir Francis Chichester. Henry Every, described as the most notorious pirate of the late 17th century, was probably born in the village of Newton Ferrers.[34] John Oxenham (1536-1580) was a lieutenant of Drake but considered a pirate by the Spanish. Thomas Morton (1576–1647?) was an avid Elizabethan outdoorsman probably born in Devon who became an attorney for The Council For New England, and built the New England fur-trading-plantation called Ma-Re Mount or Merrymount around a West Country-style Maypole, much to the displeasure of Pilgrim and Puritan colonists. Morton wrote a 1637 book *New English Canaan* about his experiences, partly in verse, and may have thereby become America's first poet to write in English.[35] Another famous mariner and Devonian was Robert Falcon Scott, the leader of the unfortunate Terra Nova Expedition to reach the geographical South Pole.[36] The poet Samuel Taylor Coleridge, the crime writer Agatha Christie and the poet Ted Hughes lived in Devon (his funeral and cremation were held there). The painter and founder of the Royal Academy, Sir Joshua Reynolds, was born in Devon.

The actor Matthew Goode was raised in Devon, and Bradley James, also an actor, was born there. The singer Joss Stone was brought up in Devon and frontman Chris Martin from the English rock group Coldplay was born there. Matt Bellamy, Dominic Howard and Christopher Wolstenholme from the English group Muse all grew up in Devon and formed the band there. Dave Hill of pop band Slade was born in Flete House which is in the South Hams district of Devon. Another famous Devonian is the model and actress Rosie Huntington-Whiteley, who was born in Plymouth and raised in Tavistock.

Trevor Francis, former Nottingham Forest and Birmingham City professional footballer was born and brought up in Plymouth. Swimmer Sharron Davies[37] was born in Plymouth. Peter Cook the satirist, writer and comedian was born in Torquay, Devon. Leicester Tigers and British Lions Rugby player Julian White MBE was born and raised in Devon and now farms a herd of pedigree South Devon beef cattle. The dog breeder John "Jack" Russell was also from Devon. Jane McGrath, who married Australian cricketer Glenn McGrath was born in Paignton, her long battle with and subsequent death from breast cancer inspired the formation of the McGrath Foundation, which is one of Australia's leading charities.

See also

- List of Lord Lieutenants of Devon
- List of High Sheriffs of Devon
- Custos Rotulorum of Devon - Keepers of the Rolls
- List of MPs for Devon constituency
- Category:Rivers of Devon
- Devonshire eggs
- List of monastic houses in Devon
- List of Sites of Special Scientific Interest in Devon
- North Devon Coast
- West Country dialects
- Circular linhay

References

[1] http://www.naturalengland.org.uk/ourwork/conservation/designatedareas/nationalparks/dartmoor.aspx|Natural England: Dartmoor retrieved 13 May 2009

[2] "William Camden, Britannia (1607) with an English translation by Philemon Holland – Danmonii" (http://www.philological.bham.ac.uk/cambrit/cornwalleng.html). The University of Birmingham. . Retrieved 30 June 2009.

[3] "Manuscript A: The Parker Chronicle" (http://asc.jebbo.co.uk/a/a-L.html). . Retrieved 29 June 2007.

[4] Davies, Norman (2000). *The Isles: A History*. p. 207. ISBN 0333692837.

[5] "Devon's Mining History and Stannary parliament" (http://users.senet.com.au/~dewnans/Devon_Stannary_History.html). users.senet.com.au. . Retrieved 29 March 2008.

[6] (http://www.swcore.co.uk/southwest.htm), South West Chamber of Rural Enterprise

[7] *In Devon, the county council estimated that 1,200 jobs would be lost in agriculture and ancillary rural industries – Hansard*, 25 April 2001 (http://www.publications.parliament.uk/pa/cm200001/cmhansrd/vo010425/debtext/10425-17.htm#column_357)

[8] Devon County Council, 2005. Tourism trends in Devon (http://www.devon.gov.uk/tourism_trends_2005.pdf).

[9] Edmonds, E. A., et al. (1975) *South-West England*; based on previous editions by H. Dewey (British Geological Survey UK Regional Geology Guide series no. 17, 4th ed.) London: HMSO ISBN 0-11-880713-7

[10] "Devon's Rocks – A Geological Guide" (http://www.devon.gov.uk/index/environmentplanning/natural_environment/geology/geology-guide.htm). Devon County Council. . Retrieved 18 May 2011.

[11] Edmonds, E. A.; McKeown, M. C.; Williams, M. (1975). "Carboniferous Rocks". *South-West England*. British Regional Geology. Dewey, H. (4th ed.). London: HMSO/British Geological Survey. p. 34. ISBN 0118807137.

[12] Dewey, Henry (1948) *British Regional Geology: South West England*, 2nd ed. London: H.M.S.O.

[13] *Whitaker's Almanack*, 1972; p. 631

[14] http://www.exmoor-nationalpark.gov.uk/index/learning_about/moor_facts.html|Exmoor National Park, National Park Facts |accessdate=10 May 2009

[15] Jones, John Pike & Kingston, J. F. (1829) *Flora Devoniensis*. 2 pts, in 1 vol. London: Longman, Rees, Orme, Brown, and Green

[16] Martin, W. Keble & Fraser, G. T. (eds.) (1939) Flora of Devon. Arbroath

[17] Paul Simons (14 May 2007). "Britain warms to the taste for home-grown olives" (http://www.timesonline.co.uk/tol/news/weather/article1785059.ece). *The Times* (UK). . Retrieved 20 September 2007.

[18] "Tories take over county council" (http://news.bbc.co.uk/1/hi/england/devon/8084708.stm). The BBC. 5 June 2009. . Retrieved 6 June 2009.

[19] "Boundary Committee publishes draft proposal for Devon" (http://www.electoralcommission.org.uk/news-and-media/news-releases/boundary-committee-news-centre/structural-reviews/boundary-committee-publishes-draft-proposal-for-devon). The Boundary Committee for

England. 7 July 2008. . Retrieved 30 July 2008.

[20] GENUKI http://genuki.cs.ncl.ac.uk/DEV/Hundreds.html

[21] Bowen, E. G. (1977) *Saints, Seaways and Settlements in the Celtic Lands*. Cardiff: University of Wales Press ISBN 0-900768-30-4

[22] "Diocese of Plymouth" (http://www.plymouth-diocese.org.uk/). . Retrieved 13 April 2009.

[23] Fox-Davies, A. C. (1915) *The Book of Public Arms*, 2nd edition, London

[24] W. C. Scott-Giles, *Civic Heraldry of England and Wales*, 2nd edition, London, 1953

[25] "A brief history of Devon's coat of arms (Devon County Council)" (http://www.devon.gov.uk/index/democracycommunities/
county_councillors/historic/brief_history.htm). Devon.gov.uk. . Retrieved 14 June 2010.

[26] "Council's designs cause logo row" (http://news.bbc.co.uk/2/hi/uk_news/england/devon/4851278.stm). BBC News. 27 March 2006. .
Retrieved 14 June 2010.

[27] "Policy and Resources Overview Scrutiny Committee Minutes, April 3, 2006" (http://www.devon.gov.uk/index/
democracycommunities/decision_making/cma/cma_document.htm?cmadoc=minutes_spr_20060403.html). Devon.gov.uk. . Retrieved 14
June 2010.

[28] "Devon Community Life – Devon gets its own flag" (http://www.bbc.co.uk/devon/community_life/features/devon_flag.shtml). BBC.
18 July 2003. . Retrieved 14 June 2010.

[29] Devon County Council Press Release, 16 October 2006 (http://www.devon.gov.uk/press_devonflagpr)

[30] "Ottery Tar Barrels" (http://www.bbc.co.uk/devon/discovering/legends/ottery_tar_barrels.shtml). BBC. . Retrieved 14 May 2008.

[31] Mason, Laura; Brown, Catherine (1999) From Bath Chaps to Bara Brith. Totnes: Prospect Books

[32] Pettigrew, Jane (2004) Afternoon Tea. Andover: Jarrold

[33] Fitzgibbon, Theodora (1972) A Taste of England: the West Country. London: J. M. Dent

[34] Marley, David F. (2010). *Pirates of the Americas*. Santa Barbara, CA: ABC-CLIO. p. 589. ISBN 9781598842012.

[35] *New English Canaan or New Canaan. Containing an abstract of New England, composed in three bookes. The first booke setting forth the
originall of the natives, their manners and customes, together with their tractable nature and love towards the English. The second booke
setting forth the naturall indowments of the country, and what staple commodities it yealdeth. The third booke setting forth, what people are
planted there, their prosperity, what remarkable accidents have happened since the first planting of it, together with their tenents and practise
of their church.* Written by Thomas Morton of Cliffords Inne gent, upon tenne yeares knowledge and experiment of the country. Amsterdam:
Jacob Stam

[36] H. G. R. King, 'Scott, Robert Falcon (1868–1912)', Oxford Dictionary of National Biography, Oxford University Press, 2004; online edn,
January 2011 accessed 21 June 2011 (http://www.oxforddnb.com/view/article/35994)

[37] "New centre to honour Plymouth Olympian Sharron Davies" (http://www.plymouth.gov.uk/newsreleases?newsid=128760). Plymouth
City Council. 14 March 2007. . Retrieved 31 August 2008.

Further reading

- Oliver, George (1846) *Monasticon Dioecesis Exoniensis: being a collection of records and instruments
illustrating the ancient conventual, collegiate, and eleemosynary foundations, in the Counties of Cornwall and
Devon, with historical notices, and a supplement, comprising a list of the dedications of churches in the Diocese,
an amended edition of the taxation of Pope Nicholas, and an abstract of the Chantry Rolls* [with supplement and
index]. Exeter: P. A. Hannaford, 1846, 1854, 1889

- Pevsner, N. (1952) *North Devon* and *South Devon* (Buildings of England). 2 vols. Penguin Books

- Stabb, John *Some Old Devon Churches: their rood screens, pulpits, fonts, etc..* 3 vols. London: Simpkin,
Marshall, Hamilton, Kent, 1908, 1911, 1916

External links

- Devon County Council (http://www.devon.gov.uk)
- BBC Devon (http://www.bbc.co.uk/devon/)
- Genuki Devon (http://www.cs.ncl.ac.uk/genuki/DEV/) Historical, geographical and genealogical
information
- The Devonshire Association (http://www.devonassoc.org.uk/history.htm), a Devon-centric equivalent of the
British Association
- Devon (http://www.dmoz.org/Regional/Europe/United_Kingdom/England/Devon/) at the Open Directory
Project

Brixham

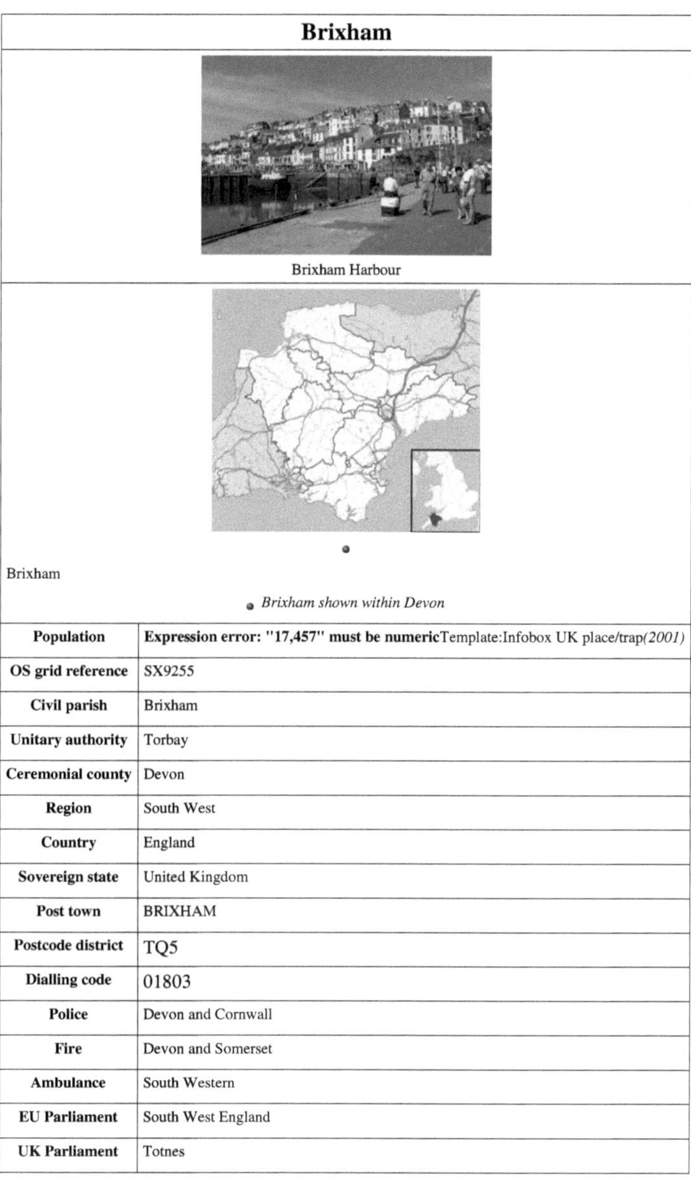

<table>
<tr><td colspan="2" align="center">Brixham</td></tr>
<tr><td colspan="2" align="center">Brixham Harbour</td></tr>
<tr><td colspan="2" align="center">Brixham
Brixham shown within Devon</td></tr>
<tr><td>Population</td><td>Expression error: "17,457" must be numericTemplate:Infobox UK place/trap(2001)</td></tr>
<tr><td>OS grid reference</td><td>SX9255</td></tr>
<tr><td>Civil parish</td><td>Brixham</td></tr>
<tr><td>Unitary authority</td><td>Torbay</td></tr>
<tr><td>Ceremonial county</td><td>Devon</td></tr>
<tr><td>Region</td><td>South West</td></tr>
<tr><td>Country</td><td>England</td></tr>
<tr><td>Sovereign state</td><td>United Kingdom</td></tr>
<tr><td>Post town</td><td>BRIXHAM</td></tr>
<tr><td>Postcode district</td><td>TQ5</td></tr>
<tr><td>Dialling code</td><td>01803</td></tr>
<tr><td>Police</td><td>Devon and Cornwall</td></tr>
<tr><td>Fire</td><td>Devon and Somerset</td></tr>
<tr><td>Ambulance</td><td>South Western</td></tr>
<tr><td>EU Parliament</td><td>South West England</td></tr>
<tr><td>UK Parliament</td><td>Totnes</td></tr>
</table>

Brixham (🔊 /ˈbrɪksəm/) is a small fishing town and civil parish in the county of Devon, in the south-west of England. Brixham is at the southern end of Torbay, across the bay from Torquay, and is a fishing port. Fishing and tourism are its major industries. At the time of the 2001 census it had a population of 17,457.[1]

It is thought that the name 'Brixham' came from Brioc's village. 'Brioc' was an old English or Brythonic personal name and '-ham' is an ancient term for village.

The town is hilly, and built around the harbour which remains in use as a dock for fishing trawlers. It has a focal tourist attraction in the replica of Sir Francis Drake's ship the *Golden Hind* that is permanently moored there.

In summer the Cowtown carnival is held, a reminder of when Brixham was two separate communities with only a marshy lane to connect them. Cowtown was the area on top of the hill where the farmers lived, while a mile away in the harbour was Fishtown, where the seamen lived. Cowtown, the St Mary's Square area, is on the road leaving Brixham to the south west, in the direction of Kingswear, upon which stands a church built on the site of a Saxon original. The local Royal British Legion club is also here. The main summer attraction in Cowtown is the Hap'nin in St. Mary's Park, an annual music event, with local bands, that takes place in mid July.

History

Although there is evidence of Ice age inhabitants here , and probable trading in the Bronze age, the first evidence of a town comes from the Saxon times. It is possible that Saxon settlement originated by sea from Hampshire in the sixth century, or overland around the year 800.[2]

Brixham was called *Briseham* in the Domesday Book.[3] Its population then was 39.[4]

Brixham was part of the former Haytor Hundred. The population was 3,671 in 1801 and 8,092 in 1901. In 1334, the town's value was assessed at one pound, twelve shillings and eightpence; by 1524, the valuation had risen to £24 and sixteen shillings. It is recorded as a borough from 1536, and a market is recorded from 1822.[5]

William Prince of Orange (afterwards King William III of Great Britain & Ireland) landed in Brixham with his mainly Dutch army, on 5 November 1688, during the Glorious Revolution, and issued his famous declaration "*The Liberties of England and The Protestant Religion I Will Maintain*". Many local people still have Dutch surnames, being direct descendants of soldiers in that army. A road leading from the harbour up a steep hill to where the Dutch made their camp, is still called Overgang, meaning 'passage' in Dutch.[6]

The coffin house reflects Brixham humour: it is coffin-shaped and when a father was asked for the hand in marriage of his daughter, he said he would 'see her in a coffin, before she wed'. The future son-in-law bought the coffin-shaped property, called it the Coffin House, and went back to the father and said 'Your wishes will be met, you will see your daughter in a coffin, the Coffin House'. Amazed by this, the father gave his blessing.[7]

The street names reflect the town's history. Pump Street is where the village pump stood. Monksbridge was a bridge built by the monks of Totnes Priory. Lichfield Drive was the route that the dead (from the Anglo-Saxon 'lich' meaning a corpse) were taken for burial at St Mary's churchyard. Salutation Mews, near the church, dates from when England was Catholic, and the salutation was to the Virgin Mary. Similarly, Laywell Road recalls Our Lady's Well. The first building seen when coming into Brixham from Paignton is the old white-boarded Toll House where all travellers had to pay a fee to keep the roads repaired.

The tower of All Saints' Church, founded in 1815, stands guard over the town. The composer of *Abide With Me*, Rev. Francis Lyte was a vicar at the church. He lived at Berry Head House, now a hotel, and when he was a very sick man, near to dying, he looked out from his garden as dusk fell over Torbay, and the words of that hymn came into his mind.

The main church is St. Mary's, about a mile from the sea. It is the third to have been on the site (which was an ancient Celtic burial ground). The original wooden Saxon church was replaced by a stone Norman church that was in its turn built over in about 1360. Many of the important townspeople are buried in the churchyard.

Many of Brixham's photogenic cottages above the harbour were originally inhabited by fishermen and their families. Near the harbour is the famous Coffin House mentioned earlier. Many of the dwellings towards Higher Brixham were built largely between the 1930s to 1970s. Several holiday camps were built in this area, for example Pontin's Wall Park and Dolphin. The Dolphin was one of the companys biggest camps. The camp closed in 1991 after fire destroyed the main entertainments building.

Brixham was served by the short Torbay and Brixham Railway from Churston. The line, opened in February 1868 to carry passengers and goods (mainly fish), was closed in May 1963 as a result of the Beeching Axe cuts. Although the former line to Brixham is deserted and overgrown, the branch line through nearby Churston is now maintained and operated as a heritage railway by a team of volunteers as the Paignton and Dartmouth Steam Railway.

Maritime

Brixham is also notable for being the town where the fishing trawler was improved in the 19th century; their distinctive sails inspired the song "Red Sails in the Sunset", which was written aboard a Brixham sailing trawler called the *Torbay Lass*.

In the Middle Ages, Brixham was the largest fishing port in the south west of England. Known as the 'Mother of Deep-Sea Fisheries', its boats helped to establish the fishing industries of Hull, Grimsby and Lowestoft. In the 1890s, there were about 300 trawling vessels in Brixham, most individually owned. The trawlers can still be seen coming in and out of the harbour, followed by flocks of seagulls. The fish market is open to the public on two special days in the summer,

Looking west across Brixham Harbour

when the finer points of catching and cooking fish are explained. The modern boats are diesel-driven, but several of the old sailing trawlers have been preserved.

Hundreds of ships have been wrecked on the rocks around the town. Brixham men have always known the dangers but even they were taken by surprise by a terrible storm that blew up on the night of 10 January 1866. The fishing boats only had sails then and could not get back into harbour because gale force winds and the high waves were against them. To make things worse, the beacon on the breakwater was swept away, and in the black darkness they could not determine their position. According to local legend, their wives brought everything they could carry, including furniture and bedding, to make a big bonfire on the quayside to guide their men home. Fifty vessels were wrecked and more than one hundred lives were lost in the storm; when dawn broke the wreckage stretched for nearly three miles up the coast.

Brixham breakwater and lighthouse

Hearing of this tragedy, the citizens of Exeter gave money to set up what became the Royal National Lifeboat Institution's Brixham Lifeboat in 1866. Now known as Torbay Lifeboat Station, it operates a Severn Class all-weather lifeboat and a D Class inshore lifeboat.[8] The crews have a history of bravery, with 52 awards for gallantry. The boathouse can be visited and memorials to the brave deeds seen; on special occasions visitors can go on board the boat. Two maroons (bangs) are the signal for the lifeboat to be launched.

Smuggling was more profitable than fishing, but if the men were caught, they were hanged. There are many legends about the local gangs and how they evaded the Revenue men. One humorous poem describes how a notorious local character, Bob Elliott ("Resurrection Bob"), could not run away because he had gout and hid in a coffin. Another villain was caught in possession but evaded capture by pretending to be the Devil, rising out of the morning mists. On another occasion when there was a cholera epidemic, some Brixham smugglers drove their cargo up from the beach in a hearse, accompanied by a bevy of supposed mourners following the cortege drawn by horses with muffled hooves.

The town's outer harbour is protected by a long breakwater, useful for sea angling. In winter this is a site for Purple Sandpiper birds. During the Second World War, a ramp and piers were built from which American servicemen left for the D-day landings.

To the south of Brixham, and sheltering the southern side of its harbour, lies the coastal headland of Berry Head with a lighthouse, Iron Age Fort and National Nature Reserve.

Military

Warships have been seen in Torbay from the days of the Vikings up until 1944 when part of the D-Day fleet sailed from here. In 1588 Brixham watched Sir Francis Drake attacking the Spanish Armada after he had (so the legend goes) finished his game of bowls on Plymouth Hoe. Today in Brixham harbour there is a full-sized replica of the ship, the *Golden Hind*, in which Drake circumnavigated the globe; visitors can go on board.

The replica of Golden Hind in Brixham harbour

For centuries, ships going down the English Channel have come into Torbay to seek refuge from the storms and to replenish food supplies. Sometimes these were merchants, taking cargoes to far away places and bringing back exotic goods and rare spices; sometimes they were carrying pilgrims, or gentlemen on the Grand Tour.

Since the days of Henry VIII Brixham has played a part in the defence of the nation. The headland known as Berry Head is now a National Nature Reserve, but it is also a military site where guns were once positioned to defend the naval ships that were re-victualling at Brixham. Twelve guns were put there during the War of American Independence, but were removed when peace came in 1783. Just ten years later, during a war with France, guns were again deployed around the town. The major position was at Berry Head, but this time fortifications were built to defend the gun positions. These can still be seen, and are now some of the best preserved Napoleonic forts in the country.

During the long series of wars against the French that began in 1689 and lasted until 1815, the Royal Navy came into Brixham to get supplies of fresh vegetables, beef and water. There might have been twenty or so of the big men-o'-war lying at anchor in Torbay, recovering from exploits of the sort described in the books about Hornblower, Bolitho or Jack Aubrey. On the harbourside towards the marina there is a grey stone building which today is the Coastguard headquarters; then, it was the King's Quay where His Majesty's vessels were provisioned. Local farmers brought vegetables to ward off scurvy, and cattle were slaughtered and their meat packed into barrels. The water came from a big reservoir situated near the crossroads in the middle of town; from there a pipeline carried it under the streets and under the harbour to the King's Quay.

Many of the well-known Admirals of the day visited Brixham. Not only Nelson, but also Lord St. Vincent, Cornwallis, Hood, Rodney and Hawke. There was also Earl Howe, who earned the nickname of Lord Torbay because he spent so much time ashore in Brixham. A notorious visitor was Napoleon Bonaparte, who, as a prisoner on *HMS Bellerophon*, spent several days off Brixham waiting to be taken to exile on St. Helena.

Battery Gardens have a military history leading back to the Napoleonic wars and the time of the Spanish Armada. The emplacements and features seen here today are those of the Second World War and are of national importance. The site, listed by English Heritage, is recognised as one of the best preserved of its kind in the UK. Of the 116 'Emergency Coastal Defence Batteries' set up in the UK in 1940, only seven remain intact.

Industrial

Apart from fishing, most of the other local industries were connected with stone. Limestone was once quarried and used to build the breakwater, for houses and roads, and was sent to Dagenham to make steel for Ford automobiles. It was also burnt in limekilns to reduce it to a powder which was spread on the land in other parts of Devon as an agricultural fertiliser. The old quarries and the limekilns can still be seen.

Another mineral found in Brixham is ochre. This gave the old fishing boats their "Red Sails in the Sunset", but the purpose was to protect the canvas from sea water. It was boiled in great caldrons, together with tar, tallow and oak bark. The latter ingredient gave its name to the barking yards which were places where the hot mixture was painted on to the sails, which were then hung up to dry. The ochre was also used to make a paint. This was invented in Brixham in about 1845 and was the first substance in the world that would stop cast iron from rusting. Other types of paint were made here as well, and the works were in existence until 1961.

There were iron mines at Brixham, and for a while they produced high quality ore but the last one closed in 1925. Most of the sites have been built over and there are now no remains of this once important industry.

Politics

On 1 April 2007, Brixham Town Council was established after a forty-year gap since Brixham Urban Council disappeared. In its first meeting the council changed its name to *Brixham Town Council* per the Local Government Act 1972 and adopted the term Chairman instead of Mayor to avoid confusion with the Torbay elected Mayor. The Council's duties are those of a standard English civil parish.

The former British Prime Minister, James Callaghan was educated partly at Furzeham Primary School.

Sport

Brixham is home to the Brixham Archers.[9] This is the biggest archery club in the bay and shoots at the Brixham Cricket Club. The Archery club was formed in 1969 and has been successful at county and national level competitions.

In 1874, Brixham Rugby union Football Club was founded and became one of the founder members of Devon RFU of which six clubs are now left. They played Rugby on Furzeham Green until 1896 when they moved their present ground to New Gate Park (now Astley Park). The club will play their league fixtures in the Southwest 1 west division of English rugby.[10]

Footballer Dan Gosling, of Newcastle United F.C., was born and raised in Brixham, and is the fourth-youngest player to have ever played for Plymouth Argyle aged 16 years and 310 days.

Transport

The railway station in Brixham on the Torbay and Brixham Railway served the town until the Brixham Line was closed in 1963. Today, the remains of the line operate between Paignton, Goodrington, Churston and Kingswear as a steam railway. There are bus services operated by Stagecoach Devon to Newton Abbot, Paignton and Torquay.

The Association of Train Operating Companies included Brixham as one of fourteen towns that, based on 2009 data, would benefit from a new railway service. This would be an extension of the First Great

Brixham Station entrance in 1964

Western service on the Riviera Line from Exmouth as far as Churston, which would then act as a railhead for Brixham. It would also serve other housing developments in the area since the opening of the steam railway, and may require the doubling of that line between Paignton and Goodrington Sands.[11]

References

[1] "Key Statistics for Torbay — May 2005" (http://www.torbay.gov.uk/appendix-b-key_statistics_for_torbay_apr.pdf) (PDF). *Census 2001*. The Consultation and Research Team, Torbay Council. p. 2. . Retrieved 2008-11-28.

[2] "About Brixham" (http://www.picturesofengland.com/England/Devon/Brixham). *Pictures Of England.com*. . Retrieved 2008-07-16.

[3] "Brixham archive" (http://www.nationalarchives.gov.uk/documentsonline/details-result.asp?Edoc_Id=7575092&queryType=1& resultcount=1). The National Archives. . Retrieved 2008-07-17.

[4] Nicholls, Richard. "Torquay, Paignton and Brixham" (http://books.google.com/?id=GO7Mhwuk6sQC& printsec=frontcover#PPA302,M1). *A History of Devonshire*. Adamant Media Corporation. p. 302. ISBN 0-543-92000-3. . Retrieved 2008-07-16.

[5] "Brixham community page" (http://www.devon.gov.uk/localstudies/100459/1.html). Devon County Council. . Retrieved 2008-07-16.

[6] "Overgang" at Lookwayup.com (http://lookwayup.com/lwu.exe/lwu/toEng?s=d&w=overgang&slang=Nld). (retrieved 17 January 2009)

[7] coffinhouse.co.uk (http://www.coffinhouse.co.uk/)

[8] Denton, Tony (2009). *Handbook 2009*. Shrewsbury: Lifeboat Enthusiasts Society. p. 59.

[9] Brixham Archers (http://www.brixhamarchers.co.uk)

[10] "RFU Competition Line-Up — 2008" (http://www.swrugby.co.uk/swcont.pdf) (PDF). *English Clubs Championship South West Leagues website*. . Retrieved 2008-07-16.

[11] "Connecting Communities - expanding access to the rail network" (http://www.atoc.org/general/ConnectingCommunitiesReport_S10. pdf). Association of Train Operating Companies. 2009. . Retrieved 2009-09-16.

External links

- Brixham (http://www.dmoz.org/Regional/Europe/United_Kingdom/England/Devon/Brixham/) at the Open Directory Project
- Brixham Town Council (http://www.torbay.gov.uk/brixhamtowncouncil)
- RNLI South Devon (http://www.rnli.org.uk/rnli_near_you/southwest/stations/TorbaySouthDevon/)
- Official Tourist Board for Torquay, Paignton and Brixham (http://www.englishriviera.co.uk)

Greenway_Estate

Greenway is an estate on the River Dart near Galmpton in Devon, England. It was first mentioned in 1493 as "Greynway", the crossing point of the Dart to Dittisham. In the late 16th century a Tudor mansion called Greenway Court was built by the Gilbert family. Greenway was the birthplace of Humphrey Gilbert. The present Georgian house was probably built in the late 18th century by Roope Harris Roope and extended by subsequent owners. It was bought by Agatha Christie and her husband Max Mallowan in 1938. The Greenway Estate was acquired by the National Trust in 1999 and it is now a Grade II* listed building. The estate is known for its large riverside gardens, which contain plants from the southern hemisphere and have been open to the public since March 2002.

Greenway House in 2010

The house was occupied by Christie and Mallowan until their deaths in 1976 and 1978 respectively. Christie's daughter Rosalind Hicks and her husband Anthony lived in the house from 1968. Rosalind Hicks died in 2004.

The garden is open to the public as is the Barn Gallery, which shows work by contemporary local artists. The house's restoration is complete and it is open to the public.

Greenway House in 2006.

Inspiration for Agatha Christie's works

Agatha Christie frequently used places familiar to her as settings for her plots. Greenway Estate and its surroundings in their entirety or in parts are described in the following novels[1] :

The restored greenhouse

* Five Little Pigs (1942)

The main house, the foot path leading from the main house to the battery overlooking the river Dart and the battery itself (where the murder occurs) are described in detail since the movements of the novel's protagonist at these locations are integral to the plot and the denouement of the murderer.

* Towards Zero (1944)

The location of the estate opposite the village of Dittisham, divided from each other by the river Dart, plays an important part for the alibi and a nightly swim of one of the suspects.

The house in July 2008, under restoration

* Dead Man's Folly (1956)

The boat house of Greenway Estate is described as the spot where the first victim is discovered, and the nearby ferry landing serves as the place where the second real murder victim is dragged into the water for death by drowning. Other places described are the greenhouse and the tennis court, where Mrs. Oliver placed real clues and red herrings for the "murder hunt". The lodge of Greenway Estate serves as the home of Amy Folliat, the former owner of Nasse House.

References

[1] John Curran: "Agatha Christie's Secret Notebooks - Fifty years of mystery in the making", Paperback edition, HarperCollinsPublishers 2010, ISBN 978-0-00-731057-9

* Campbell, Sophie (February 24, 2009). "Agatha Christie's home Greenway opens to the Devon public" (http:// www.telegraph.co.uk/travel/artsandculture/4797440/ Agatha-Christies-home-Greenway-opens-to-the-Devon-public.html). London: www.telegraph.co.uk. Retrieved 2009-02-24.

- "National Trust - Greenway - History" (http://www.nationaltrust.
  org.uk/main/w-vh/w-visits/w-findaplace/w-greenway/
  w-greenway-history.htm). www.nationaltrust.org.uk. Retrieved
  2008-02-04.
- Norway, Arthur Hamilton (1897) *Highways and Byways in Devon
  and Cornwall*. London: Macmillan; p. 100

The Vinery

Further reading

- *Greenway, Devon*. London: National Trust, 2003; revised 2006, 24 pages. ISBN 978-1-84359-079-8

External links

- Greenway information at the National Trust (http://nationaltrust.org.uk/greenway)
- Images of Greenway (http://www.ntprints.com/search.php?page=1&numperpage=12&idx=0&
  keywords=greenway&ref=wiki&ad=nt002) on the Official National Trust Print Website
- Video tour of Greenway (http://www.bbc.co.uk/archive/agatha_christie/12516.shtml) from a BBC newscast

Listed_building

A **listed building** in the United Kingdom is a building that has been placed on the **Statutory List of Buildings of Special Architectural or Historic Interest**. It is a widely used status, applied to around half a million buildings.

A listed building may not be demolished, extended or altered without special permission from the local planning authority (which typically consults the relevant central government agency, particularly for significant alterations to the more notable listed buildings). Exemption from secular listed building control is provided for some buildings in current use for worship but only in cases where the relevant religious organisation operates its own equivalent permissions procedure. Owners of listed buildings are, in some circumstances, compelled to repair and maintain them and can face criminal prosecution if they fail to do so or if they perform unauthorised alterations.

The listing procedure allows for buildings to be removed from the list if the listing is shown to be in error.[1]

Although most structures appearing on the lists are buildings, other structures such as bridges, monuments, sculptures, war memorials, and even milestones and mileposts and the Beatles' Abbey Road pedestrian crossing[2] are also listed. Ancient, military and uninhabited structures (such as Stonehenge) are sometimes instead classified as Scheduled Ancient Monuments and protected by much older legislation whilst cultural landscapes such as parks and gardens are currently "listed" on a non-statutory basis. Slightly different systems operate in each area of the United Kingdom, though the basic principles of listing are the same.

The Forth Bridge, designed by Sir Benjamin Baker and Sir John Fowler, opened in 1890, and now owned by Network Rail, is designated as a Category A listed building by Historic Scotland.

Background

Although a limited number of 'ancient monuments' were given protection under the Ancient Monuments Protection Act 1882,[3] there was reluctance to restrict the owners of occupied buildings in what they could do to their property. It was the damage to buildings caused by German bombing during World War II that prompted the first listing of buildings that were deemed to be of particular architectural merit.[4] 300 members of the Royal Institute of British Architects and the Society for the Protection of Ancient Buildings were dispatched to prepare the list under the supervision of the Inspectorate of Ancient Monuments, with funding from the Treasury.[5] The listings were used as a means of determining whether a particular building should be rebuilt if it was damaged by bombing,[4] with varying degrees of success.[5]

The basis of the current more comprehensive listing process was developed from the wartime system and was enacted by a provision in the Town and Country Planning Act 1947 covering England and Wales, and the Town and Country Planning (Scotland) Act 1947 covering Scotland. Listing was first introduced into Northern Ireland under the Planning (Northern Ireland) Order 1972. The listing process has since developed slightly differently in each part of the UK.

Heritage protection

In the UK, the process of protecting the built historic environment (i.e. getting a heritage asset legally protected) is called 'designation'. To complicate things, several different terms are used because the processes use separate legislation: buildings are 'listed'; ancient monuments are 'scheduled', wrecks are 'protected', and battlefields, gardens and parks are 'registered'. A heritage asset is a part of the historic environment that is valued because of its historic, archaeological, architectural or artistic interest.[6] Only some of these are judged to be important enough to have extra legal protection through designation. However, buildings that are not formally listed, but still judged as being of heritage interest are still regarded as being a material consideration in the planning process.[7]

As a very rough guide, listed buildings generally have substantial remains that are visible above the ground whereas ancient monuments are (mostly) below the ground and/or unoccupied.[8]

What can be listed

Almost anything can be listed – it does not have to be a building. Buildings and structures of special historic interest come in a wide variety of forms and types, ranging from telephone boxes and road signs, to castles. English Heritage has created twenty broad categories of structures, and published selection guides for each one to aid with assessing buildings and structures. These include historical overviews and describe the special considerations for listing each category.[9] [10] Neither Historic Scotland nor Cadw appear to have published comparable guidelines for particular categories (as of June 2011) although both organisations produce guidance for owners.

How to apply for listing or delisting

In England, to have a building considered for listing or de-listing, the process is to submit an application form online to English Heritage. The applicant does not need to be the owner of the building in order to apply for it to be listed.[10] Full information including application form guidance notes are on the English Heritage website. English Heritage assesses buildings put forward for listing or de-listing and provides advice to the Secretary of State on the architectural and historic interest. The Secretary of State, who may seek additional advice from others, then decides whether or not to list or de-list the building.

In Wales, applications are made using a form obtained from the relevant local authority.[11] There is no provision for consent to be granted in outline. When a local authority is disposed to grant listed building consent, it must first notify the National Assembly (i.e. Cadw) of the application. If the planning authority decides to refuse consent, it may do so without any reference to Cadw.

In Scotland, applications are made using a form obtained from Historic Scotland. After consultation with the local planning authority, the owner, where possible, and an independent third party, Historic Scotland will then make a recommendation on behalf of the Scottish Ministers.[12]

England and Wales

The legislation relevant to listing

In England and Wales the authority for listing is granted to the Secretary of State by the Planning (Listed Buildings and Conservation Areas) Act 1990. Listed buildings in danger of decay are listed on the English Heritage 'Heritage at Risk' Register.

In 1980 there was public outcry at the sudden destruction of the art deco Firestone Factory (Wallis, Gilbert and Partners, 1928–29), which was demolished over the August bank holiday weekend by its owners Trafalgar House who had been told that it was likely to be 'spot-listed' a few days later,[13] and the Government undertook to review arrangements for listing buildings.[14] After the Firestone demolition, the Secretary of State for the Environment Michael Heseltine also initiated a complete re-survey of buildings to ensure there was nothing which merited preservation and had been missed off the lists.[15]

In England, the Department for Culture, Media and Sport (DCMS) works with English Heritage (an agency of the DCMS), and other government departments, e.g. Department for Communities and Local Government (DCLG) and the Department for the Environment, Food and Rural Affairs (DEFRA) to deliver the government policy on the protection to historic buildings and other heritage assets. The decision about whether or not to list a building is made by the Secretary of State, although the process is administered in England by English Heritage.[16] In Wales (where it is a devolved issue) it is administered by Cadw on behalf of the National Assembly for Wales[17] and in Scotland it is administered by Historic Scotland on behalf of the Scottish Ministers.[18]

Heritage protection reform legislation in England

There have been several attempts to simplify the heritage planning process for listed buildings in England, which has still (at the time of writing in May 2011) to reach a conclusion.[19]

The review process was started in 2000 by Alan Howarth, then minister at the Department for Culture, Media and Sport (DCMS). The outcome was the paper 'The Power of Place' in 2000[20] followed by the subsequent policy document 'The Historic Environment: A Force for Our Future' published by the DCMS and the Department of the Environment, Transport and the Regions (DTLR) in December 2001.[21] The launch of the Government's Heritage Protection Reform (HPR) report in July 2003 by the DCMS entitled: 'Protecting our historic environment: Making the system work better,'[22] asked questions about how the current designation systems could be improved. The HPR decision report 'Review of Heritage Protection: The Way Forward' green paper published in June 2004 by the DCMS committed the UK government and English Heritage to a process of reform including a review of the criteria used for listing buildings.

The Government also began a process of consultation on changes to Planning Policy Guidance 15 (PPG 15) relating to the principles of selection for listing buildings in England. After several years of consultation with heritage groups, charities, planning authorities and English Heritage, this eventually resulted in the publication of Planning Publication Statement 5 'Planning for the Historic Environment' in March 2010 by the DCLG. This replaced PPG15 and sets out the government's national policies on the conservation of the historic environment for the England.[7] PPS5 is supported by a Practice Guide, endorsed by the DCLG, the DCMS, and English Heritage[7] which describes how to apply the policies stated in PPS5.

The government's White Paper 'Heritage Protection for the 21st Century' published on 8 March 2007 offered a commitment to sharing the understanding of the historic environment and more openness in the process of designation.[23]

In 2008, a draft Heritage Protection Bill[24] was subject to pre-legislative scrutiny before its passage through UK Parliament. In the event, the legislation was abandoned despite strong cross-party support, to make room in the parliamentary legislative programme for measures to deal with the credit crunch.[25] though it may be revived in future. The proposal was that the existing registers of buildings, parks and gardens, archaeology and battlefields, maritime wrecks, and World Heritage Sites be merged into a single online register which will "explain what is special and why". English Heritage would become directly responsible for identifying historic assets in England and there would be wider consultation with the public and asset owners, and new rights of appeal. There would have been streamlined systems for granting consent for work on historic assets.[26]

Categories of listed building

There are three types of listed status for buildings in England and Wales:[27]

- Grade I: buildings of exceptional interest,
- Grade II*: particularly important buildings of more than special interest.
- Grade II: buildings that are of special interest, warranting every effort to preserve them.[28]

There was formerly a non-statutory Grade III, which was abolished in 1970.[29] Additionally, Grades A, B and C were used mainly for Anglican churches in use – these correspond approximately to Grades I, II* and II. These grades were used mainly before 1977, although a few buildings are still listed using these grades.

Listed buildings account for about 2% of English building stock.[30] In March 2010, there were approximately 374,000 list entries[16] of which 92% were Grade II, 5.5% were Grade II*, and 2.5% were Grade I.[31] Places of worship play an important role in the UK's architectural heritage. England alone has 14,500 listed places of worship (4,000 Grade I, 4,500 Grade II* and 6,000 Grade II). In fact, 45% of all Grade I listed buildings are places of worship.[32]

There are estimated to be about 500,000 actual buildings listed, as listing entries can apply to more than one building.

Statutory Criteria for listing

In order to be listed, a building must meet various criteria.[33] The criteria for listing include architectural interest, historic interest and close historical associations with significant people or events. Buildings which are not individually noteworthy may still be listed if they form part of a group that is – for example, all the buildings in a square. This is called 'group value'. Sometimes large areas comprising many buildings may not justify listing but are given the looser protection of designation as a conservation area.

The criteria include:

- **Age and rarity:** The older a building is, the more likely it is to be listed. All buildings erected before 1700 "which contain a significant proportion of their original fabric" will be listed. Most buildings built between 1700–1840 are listed. After 1840 more selection is exercised and "particularly careful selection" is applied after 1945. Buildings less than 30 years old are rarely listed unless they are of outstanding quality and under threat.
- **Aesthetic merits:** i.e. the appearance of a buildings. However, buildings that have little visual appeal may be listed on grounds of representing particular aspects of social or economic history.
- **Selectivity:** where a large number of buildings of a similar type survive, the policy is only to list those which are the most representative or significant examples.
- **National interest:** significant or distinctive regional buildings e.g. those that represent a nationally important but localised industry
- **State of repair:** this is **not** deemed to be a relevant consideration for listing. A building can be listed regardless of its state of repair.[33]

Additionally:

- Any buildings or structures constructed before 1 July 1948 which fall within the curtilage of a listed building are treated as part of the listed building.[34]

- The effect of a proposed development on the setting of a listed building is a material consideration in determining a planning application. Setting is defined as "the surroundings in which a heritage is experienced".[7]

Although the decision to list a building may be made on the basis of the architectural or historic interest of one small part of the building, the listing protection nevertheless applies to the whole building. Listing applies not just to the exterior fabric of the building itself, but also to the interior, fixtures, fittings, and objects within the curtilage of the building even if they are not fixed.[35]

De-listing is possible but rare in practice. One example is the November 30, 2001 de-listing of North Corporation Primary School, Liverpool.

Emergency listing

In an emergency, the local planning authority can serve a temporary listed "building preservation notice", if a building is in danger of demolition or alteration in such a way that might affect its historic character.[35] This remains in force for 6 months until the Secretary of State decides whether or not to formally list the building.

Certificates of immunity

If planning permission is being sought or has been obtained in England, anyone can ask the Secretary of State to issue a Certificate of Immunity (CoI) in respect of a particular building. CoIs give certainty to developers proposing works that will affect buildings that may be eligible for listing. To apply for a Certificate of Immunity, it is necessary to submit an application form. Guidance notes are available on English Heritage website.

Altering a listed building

In England and Wales, the management of listed buildings is the responsibility of local planning authorities and the Department for Communities and Local Government (i.e. not DCMS which originally listed the building). There is a general principle that listed buildings are put to 'appropriate and viable use' and recognise that this may involve the re-use and modification of the building.[7] However, listed buildings cannot be modified without first obtaining Listed Building Consent through the relevant local planning authority[36]

Carrying out unauthorised works to a listed building is a criminal offence and owners can be prosecuted. A planning authority can also insist that all work undertaken without consent be reversed at the owner's expense.

Examples of Grade I listed buildings

See also Category:Grade I listed buildings for examples of such buildings across England and Wales

- Clifton Suspension Bridge, Bristol
- The Palace of Westminster, London
- Royal Albert Hall, London
- York Minster, York
- Blackpool Tower, Blackpool
- Leeds Town Hall, Leeds
- Albert Dock, Liverpool
- Curzon Street railway station, Birmingham
- Warwick Castle, Warwick (Warwickshire)
- Montacute House, South Somerset
- Dock Tower, Grimsby
- Royal Festival Hall, London (first Grade I-listed postwar building)
- Lilford Hall, Northamptonshire

Buckingham Palace, the official London residence of the British monarch, listed Grade I.

- Pontcysyllte Aqueduct, North Wales
- Lloyd's building, City of London
- Birmingham Town Hall, Birmingham

Examples of Grade II* listed buildings

See also Category:Grade II listed buildings for examples of such buildings across England and Wales*

- Bank Hall, Bretherton
- Battersea Power Station, London
- Coliseum Theatre, London
- Middlesbrough Transporter Bridge, Middlesbrough
- Shibden Hall, Calderdale
- Manchester Town Hall extension, Manchester
- St John's Jerusalem, Kent
- Trellick Tower, London
- Birmingham Council House, Birmingham

The Bank Hall mansion house is a Grade II* listed building, due to the 17th century clock tower, which features an original oak cantilevered staircase.

Examples of Grade II listed buildings

See also Category:Grade II listed buildings for examples of such buildings across England and Wales

- Alexandra Palace, London
- Broomhill Pool, Ipswich
- BT Tower, London
- Whitechapel Bell Foundry, London
- Birmingham Back to Backs, Birmingham

Mixed designations

The Johnny Haynes stand at Craven Cottage is a Grade II* listed building.

- In 2002, there were 80 seaside piers in England that were listed, variously at Grades I, II* and II.
- Golden Lane Estate, City of London, is an example of a site which includes buildings of different Grades, II and II*
- Cobham Park, Kent, is a Listed Landscape (Humphry Repton and older landscape) containing both Grade I structures (Cobham Hall and Darnley Mausoleum) and Grade II structures (ornamental dairy etc.) as well as a Scheduled Ancient Monument (a buried Roman villa).
- West Norwood Cemetery is the first-ever Gothic-designed cemetery and crematorium which contains 65 structures of Grade II or II*, mainly sepulchral monuments but also boundary structures and mausolea. Additionally it is listed Grade II* on the Register of Parks and Gardens.
- Derwent Valley Mills includes 838 listed buildings, made up of 16 Grade I, 42 Grade II*, and 780 Grade II. A further nine structures are Scheduled Ancient Monuments.

Locally listed buildings

Many councils, for example, Birmingham City Council, maintain a list of *locally listed buildings* as separate to the statutory list (and in addition to it). There is no statutory protection of a building or object on the local list. Councils hope that owners will recognise the merits of their properties and keep them unaltered if at all possible.

These grades are used by Birmingham:

Grade A

> This is of statutory list quality. To be the subject of notification to English Heritage and/or the serving of a Building Preservation Notice if imminently threatened.

Grade B

> Important in the city wide architectural or local street scene context, warranting positive efforts to ensure retention.

Grade C

> Of significance in the local historical/vernacular context, including industrial archaeological features, and worthy of retention.

Northern Ireland

Listing began later in Northern Ireland than in the rest of the UK: the first provision for listing was contained in the Planning (Northern Ireland) Order 1972; and the current legislative basis for listing is the Planning (Northern Ireland) Order 1991.[38] Under Article 42 of the Order, the Department of the Environment of the Northern Ireland Executive is required to compile lists of buildings of "special architectural or historic interest". The responsibility for the listing process rests with the Northern Ireland Environment Agency (NIEA), an executive agency within the Department of the Environment.[38]

The Grade-A-listed Mussenden Temple, County Londonderry[37]

Following the introduction of listing, an initial survey of Northern Ireland's building stock was begun in 1974.[39] By the time of the completion of this First Survey in 1994, the listing process had developed considerably, and it was therefore decided to embark upon a Second Survey to update and cross-check the original information. As of April 2010, the Second Survey had been completed for 147 of Northern Ireland's 547 council wards, and completion is anticipated by 2016.[39] Information gathered during this survey, relating to both listed and unlisted buildings, is entered into the publicly-accessible Northern Ireland Buildings Database.[39] A range of listing criteria, which aim to define architectural and historic interest, have been developed by the NIEA, and are used to determine whether or not to list a building.[38] Listed building consent must be obtained from local authorities prior to any alteration to a listed structure.[40]

The scheme of listing is as follows:

- **Grade A**: "buildings of greatest importance to Northern Ireland including both outstanding architectural set-pieces and the least altered examples of each representative style, period and type."[41]
- **Grade B+**: "buildings which might have merited grade A status but for detracting features such as an incomplete design, lower quality additions or alterations. Also included are buildings that because of exceptional features, interiors or environmental qualities are clearly above the general standard set by grade B buildings. A building may merit listing as grade B+ where its historic importance is greater than a similar building listed as grade B."[41]

- **Grade B**: "buildings of local importance and good examples of a particular period or style. A degree of alteration or imperfection of design may be acceptable."[41]

There are approximately 8,500 listed buildings in Northern Ireland, representing 2% of the total building stock.[38] Of these, around 200 are listed at Grade A, 400 at Grade B+, and the remainder at Grade B.[41] Since 1987, buildings within Grade B are separated into Grade B1, which is applied to buildings that qualify on a wider range of attributes, and Grade B2, which is applied to buildings with a narrower range of qualifying features.[41]

Examples of Grade A listed buildings

- Bangor Abbey, County Down[42]
- Grand Opera House, Belfast[43]

Examples of Grade B+ listed buildings

- Dundarave House, County Antrim[44]
- Necarne, County Fermanagh[45]

Examples of Grade B1 listed buildings

- Campbell College, Belfast[46]
- Linen Hall Library, Belfast[47]

Scotland

In Scotland, listing was begun by a provision in the Town and Country Planning (Scotland) Act 1947, and the current legislative basis for listing is the Planning (Listed Buildings and Conservation Areas) (Scotland) Act 1997.[49] As with other matters regarding planning, conservation is a power devolved to the Scottish Parliament and the Scottish Government. The authority for listing rests with Historic Scotland, an executive agency of the Scottish Government, which inherited this role from the Scottish Development Department in 1991. Listed building consent must be obtained from local authorities prior to any alteration to a listed structure.[49]

The National Gallery of Scotland, Edinburgh, designed by William Henry Playfair, is a Category A listed building[48]

The scheme for classifying buildings is:

- **Category A**: "buildings of national or international importance, either architectural or historic, or fine little-altered examples of some particular period, style or building type."[50]
- **Category B**: "buildings of regional or more than local importance, or major examples of some particular period, style or building type which may have been altered."[50]
- **Category C(S)**: "buildings of local importance, lesser examples of any period, style, or building type, as originally constructed or moderately altered; and simple traditional buildings which group well with others in categories A and B."[50]

There are approximately 47,400 listed buildings in Scotland. Of these, around 8 percent (some 3,800) are Category A, and 51 percent (24,000) are Category B, with the rest listed at Category C(s).[51]

Examples of Category A listed buildings

- Craigellachie Bridge, Moray[53]
- Glasgow City Chambers, Glasgow[54]
- Palace of Holyroodhouse, Edinburgh[55]
- Ravelston Garden, Edinburgh[56]

Examples of Category B listed buildings

- Harbourmaster's House, Dysart, Fife[57]
- National War Museum of Scotland, within Edinburgh Castle[58]
- Sabhal Mòr Ostaig, Isle of Skye[59]

The main stand of Ibrox Stadium, the home of Rangers F.C., is a Category B listed building[52]

Examples of Category C(S) listed buildings

- St John's Cathedral, Oban, Argyll[60]
- The Belmont Picturehouse, Aberdeen[61]
- Craigend Castle, East Dunbartonshire[62]

How to find a listed building

Although the 2008 draft legislation was abandoned, English Heritage published a single list of all designated heritage assets within England in 2011.[63] The National Heritage List for England is an online searchable database which includes 400,000 (most but not all) of England's listed buildings, scheduled monuments, registered parks and gardens, protected historic wrecks and registered battlefields in one place. The legislative frameworks for each type of historic asset remains unchanged (2011).[64]

In Scotland, the national dataset of listed buildings and other heritage assets can be searched online via Historic Scotland,[65] or through the map database Pastmap.[66]

To find a listed building in Wales, it is necessary to contact the appropriate local authority or Cadw. Also British Listed Buildings (website) [67] has sections on England, Wales and Scotland. It can be searched either by browsing for listed buildings by country, county and parish/locality, or by keyword search or via the online map. Not all buildings have photographs, as it is run on a volunteer basis.

The Northern Ireland Buildings Database contains details of all listed buildings in Northern Ireland.[68]

A photographic library of English listed buildings was started in 1999 as a snap shot of buildings listed at the turn of the millennium. This is not an up-to-date record of all listed buildings in England - the listing status and descriptions are only correct as at February 2001.[69] The photographs were taken between 1999 and 2008. It is maintained by the English Heritage archive at the Images of England project website. The National Heritage List for England contains the up-to-date list of listed buildings.[64]

Listed buildings in danger of being lost through damage or decay in England started to be recorded by survey in 1991.[70] This was extended in 1998 with the publication of English Heritage's 'Buildings at Risk Register' which surveyed Grade I and Grade II* buildings. In 2008 this survey was renamed 'Heritage at Risk' and extended to include all listed buildings, scheduled monuments, registered parks and gardens, registered battlefields, protected wreck sites and conservation areas.[71] The register is complied by survey using information from local authorities, official and voluntary heritage groups and the general public. It is possible to search this list online.[72]

In Scotland, a buildings at risk register was started in 1990 by the Royal Commission on the Ancient and Historical Monuments of Scotland (RCAHMS)in response to similar concerns at the number of listed buildings that were vacant and in disrepair. RCAHMS maintain the register on behalf of Historic Scotland,[73] and provides information on properties of architectural or historic merit throughout the country that are considered to be at risk.

In Wales, at risk registers of listed buildings are complied by local planning authorities and CADW produced a report in 2009.[74] The Royal Commission on the Ancient and Historical Monuments of Wales's (RCAHMW) Emergency Buildings Recording team is responsible for surveying historic buildings threatened with destruction, substantial alteration, or serious decay.

Equivalent status outside the United Kingdom

- ▊ ▊ France - Monument historique ("historical monument")
- ▬ Germany - Deutsche Stiftung Denkmalschutz ("German Foundation for Monument Protection") and National Heritage Sites (Kulturdenkmal ("cultural monument"))
- ☆ Hong Kong - Historic Building, see List of Grade I historic buildings in Hong Kong, List of Grade II historic buildings in Hong Kong and List of Grade III historic buildings in Hong Kong
- ▬ Netherlands - Rijksmonument ("national monument")
- ▨▨ New Zealand - New Zealand Historic Places Trust
- ▌▌ Sweden - Byggnadsminne ("monument", "memory building")
- ▨▨ United States - National Register of Historic Places and National Historic Landmark

For other countries' equivalents see List of heritage registers.

See also

- Cadw
- English Heritage
- Historic Scotland
- National Monuments Record (England)
- Conservation area
- Images of England
- Grade I listed buildings in England by county
- Grade II* listed buildings in England by county
- Grade II listed buildings in England by county
- Register of Historic Parks and Gardens of special historic interest in England
- Scheduled Ancient Monument
- Site of Special Scientific Interest
- Tree preservation order

References

[1] "Listing FAQs" (http://www.english-heritage.org.uk/caring/listing/listed-buildings/listing-faqs/). English Heritage. . Retrieved 24 May 2011.

[2] "The unusual buildings granted listed status" (http://www.telegraph.co.uk/news/uknews/8562655/ The-unusual-buildings-granted-listed-status.html). *Daily Telegraph.* 8 June 2011. . Retrieved 9 June 2011.

[3] "Preserving historic sites and buildings" (http://www.parliament.uk/about/livingheritage/transformingsociety/landscape/overview/ historicsites.cfm). Parliament.uk. . Retrieved 2010-08-27.

[4] Listed buildings (http://www.victoriansociety.org.uk/advice/listed-buildings/), The Victorian Society

[5] Targets of enemy bombers and our own demolition men (http://www.independent.co.uk/arts-entertainment/art/news/ targets-of-enemy-bombers-and-our-own-demolition-men-1598384.html), The Independent

[6] "Protecting the Historic Environment" (http://www.culture.gov.uk/what_we_do/historic_environment/6258.aspx). *DCMS.* . Retrieved 7 June 2011.

[7] "Planning policy Statement 5 'Planning for the Historic Environment'" (http://www.communities.gov.uk/publications/ planningandbuilding/pps5). *Dept of Communities and Local Government.* March 2010. . Retrieved 7June 2011.

[8] "Planning (Listed Buildings and Conservation Areas) Act 1990" (http://www.legislation.gov.uk/ukpga/1990/9/contents). *UK Government*. June 1990. . Retrieved 7 June 2011.

[9] "Selection Guildlines" (http://www.english-heritage.org.uk/caring/listing/criteria-for-protection/selection-guidelines/). *English HEritage*. . Retrieved 8 June 2011.

[10] "Listed Buildings FAQs" (http://www.wiltshire.gov.uk/planninganddevelopment/ourplanningservices/conservationhistoricenv/conservationlistedbldgs.htm). *Wiltshire Council*. . Retrieved 8 June 2011.

[11] "Listed Building Consent" (http://www.cadw.wales.gov.uk/upload/resourcepool/LBC-e1895.html). *Cadw*. . Retrieved 8 June 2011.

[12] "The Listing PRocess" (http://www.historic-scotland.gov.uk/index/heritage/historicandlistedbuildings/the-listing-process.htm). *Historic Scotland*. . Retrieved 8 June 2011.

[13] John Witherow, "No listing of Hoover factory", *The Times*, 1 September 1980, p. 4.

[14] John Young, "A notable dozen are added to the nation's listed buildings", *The Times*, 15 October 1980, p. 4.

[15] Charles Knevitt, "Protecting palaces and pillarboxes", *The Times*, 3 June 1985, p. 8.

[16] "Listing Buildings" (http://www.culture.gov.uk/what_we_do/historic_environment/3330.aspx). *DCMS Accessdate=6 June 2011*. .

[17] "Listed Buildings in Wales" (http://www.cadw.wales.gov.uk/upload/resourcepool/LBC-e1895.html). *Cadw*. . Retrieved 7 June 2011.

[18] "The Listing Process" (http://www.historic-scotland.gov.uk/index/heritage/historicandlistedbuildings/the-listing-process.htm). *Historic Scotland*. . Retrieved 7 June 2011.

[19] John Sharland (2006). "Listed Buildings and the Historic Environment - A Critique of the Government's Review of Heritage Policy'" (http://www.sharpepritchard.co.uk/). . Retrieved 23 May 2011.

[20] "The Power of Place" (http://www.english-heritage.org.uk/publications/power-of-place/powerofplacecover1.pdf). 2000. . Retrieved 23 May 2011.

[21] "The Historic Environment: A Force for our Future" (http://webarchive.nationalarchives.gov.uk/+/http://www.culture.gov.uk/reference_library/publications/4667.aspx). 2001. . Retrieved 7 June 2011.

[22] "Protecting our Historic Environment: Making the System Work Better" (http://webarchive.nationalarchives.gov.uk/+/http://www.culture.gov.uk/reference_library/consultations/1168.aspx). 2003. . Retrieved 7 June 2011.

[23] "Selection Guidelines" (http://www.english-heritage.org.uk/caring/listing/criteria-for-protection/selection-guidelines/). *English Heritage*. . Retrieved 7 June 2011.

[24] "Draft Heritage Protection Bill" (http://www.official-documents.gov.uk/document/cm73/7349/7349.pdf). *DCSM*. April 2009. . Retrieved * June 2011.

[25] Roger Mascall (2009-12-18). "The Heritage Protection Bill Fundamental reform for England and Wales?" (http://www.buildingconservation.com/articles/heritagepbill/heritagepbill.htm). . Retrieved 7 June 2011.

[26] English Heritage (http://www.english-heritage.org.uk/server/show/ConWebDoc.13381)

[27] "Principles of Selection for Listing Buildings" (http://www.culture.gov.uk/images/publications/Principles_Selection_Listing.pdf). *DCMS*. March 2010. . Retrieved 24 May 2011.

[28] "Listed Buildings" (http://www.english-heritage.org.uk/caring/listing/listed-buildings/). English Heritage. . Retrieved 24 May 2011.

[29] "About Listed Buildings" (http://www.heritage.co.uk/apavilions/glstb.html). heritage.co.uk. .

[30] "Heritage at Risk Report" (http://www.english-heritage.org.uk/publications/har-2010-report/HAR-report-2010.pdf). *English Heritage*. July 2010. . Retrieved 6 June 2011.

[31] "Listed Buildings" (http://www.english-heritage.org.uk/caring/listing/listed-buildings/). *English Heritage*. . Retrieved 7 June 2011.

[32] "Caring for Places of Worship" (http://www.brin.ac.uk/news/?tag=places-of-worship). *British Religion in Numbers*. . Retrieved 24 May 2011.

[33] "Principles of Selection for Listing Buildings" (http://www.culture.gov.uk/images/publications/Principles_Selection_Listing.pdf). DCMS March 2010. . Retrieved 25 May 2011.

[34] Planning (Listed Buildings and Conservation Areas) Act 1990, Part 1, Chapter 1, Section 5(a) (http://www.opsi.gov.uk/acts/acts1990/ukpga_19900009_en_2#pt1-ch1-l1g1).

[35] section 1.6 "Planning (Listed Buildings and Conservation Areas) Act 1990:Listing of buildings of special architectural or historic interest" (http://www.legislation.gov.uk/ukpga/1990/9/part/I/chapter/1). *UK Government*. section 1.6. Retrieved 8 June 2011.

[36] "What can I do with my listed building" (http://www.english-heritage.org.uk/your-property/planning-advice/what-can-i-do-with-my-listed-building/). *English Heritage*. . Retrieved 8 June 2011.

[37] Mussenden Temple: Historic Building Details (http://www.ni-environment.gov.uk/content-databases-buildview?id=8766&js=true). *Northern Ireland Buildings Database*. Northern Ireland Environment Agency. Retrieved 2010-07-06.

[38] "Criteria for Listing: A consultation on proposed revisions to Annex C of Planning Policy Statement 6" (http://www.ni-environment.gov.uk/criteria_for_listing_consultation_2010.pdf). Northern Ireland Environment Agency. 2010. . Retrieved 2010-07-06.

[39] "Second Survey" (http://www.ni-environment.gov.uk/built-home/recording/historic_buildings_r/second_survey.htm). Northern Ireland Environment Agency. . Retrieved 2010-07-06.

[40] "Planning Policy Statement 6: Planning, Archaeology, and the Built Heritage" (http://www.planningni.gov.uk/index/policy/policy_publications/planning_statements/pps06-archaeology-built-heritage.pdf). Planning Service. March 1999. p. 22. . Retrieved 2010-07-06.

[41] "Planning Policy Statement 6: Planning, Archaeology, and the Built Heritage" (http://www.planningni.gov.uk/index/policy/policy_publications/planning_statements/pps06-archaeology-built-heritage.pdf). Planning Service. March 1999. pp. 48–49. . Retrieved

2010-07-06.

[42] Bangor Abbey Parish Church: Historic Building Details (http://www.ni-environment.gov.uk/content-databases-buildview?id=6306& js=true). *Northern Ireland Buildings Database*. Northern Ireland Environment Agency. Retrieved 2010-07-06.

[43] Grand Opera House: Historic Building Details (http://www.ni-environment.gov.uk/content-databases-buildview?id=8650&js=true). *Northern Ireland Buildings Database*. Northern Ireland Environment Agency. Retrieved 2010-07-06.

[44] Dundarave: Historic Building Details (http://www.ni-environment.gov.uk/content-databases-buildview?id=1203&js=true). *Northern Ireland Buildings Database*. Northern Ireland Environment Agency. Retrieved 2010-07-06.

[45] Necarne Castle: Historic Building Details (http://www.ni-environment.gov.uk/content-databases-buildview?id=1548&js=true). *Northern Ireland Buildings Database*. Northern Ireland Environment Agency. Retrieved 2010-07-06.

[46] Campbell College: Historic Building Details (http://www.ni-environment.gov.uk/content-databases-buildview?id=2764&js=true). *Northern Ireland Buildings Database*. Northern Ireland Environment Agency. Retrieved 2010-07-06.

[47] Linen Hall Library: Historic Building Details (http://www.ni-environment.gov.uk/content-databases-buildview?id=4184&js=true). *Northern Ireland Buildings Database*. Northern Ireland Environment Agency. Retrieved 2010-07-06.

[48] "National Gallery of Scotland: Listed Building Report" (http://hsewsf.sedsh.gov.uk/hslive/hsstart?P_HBNUM=27679). Historic Scotland. . Retrieved 2010-07-06.

[49] *Scottish Historic Envronment Policy* (http://www.historic-scotland.gov.uk/shep.pdf). Historic Scotland. October 2008. pp. 24–25. ISBN 978-1-84917-002-4. . Retrieved 2010-07-06.

[50] "What is Listing?" (http://www.historic-scotland.gov.uk/index/heritage/historicandlistedbuildings/listing.htm). Historic Scotland. . Retrieved 2010-07-06.

[51] *Guide to the Protection of Scotland's Listed Buildings* (http://www.historic-scotland.gov.uk/scotlands-listed-buildings.pdf). Historic Scotland. 2009. p. 17. ISBN 978-1-84917-013-0. . Retrieved 2010-07-06.

[52] "Ibrox Stadium: Listed Building Report" (http://hsewsf.sedsh.gov.uk/hslive/hsstart?P_HBNUM=33383). Historic Scotland. . Retrieved 2010-07-06.

[53] "Craigellachie, Old Bridge: Listed Building Report" (http://hsewsf.sedsh.gov.uk/hslive/hsstart?P_HBNUM=2357). Historic Scotland. . Retrieved 2010-07-06.

[54] "Glasgow City Chambers: Listed Building Report" (http://hsewsf.sedsh.gov.uk/hslive/hsstart?P_HBNUM=32691). Historic Scotland. . Retrieved 2010-07-06.

[55] "Palace of Holyroodhouse: Listed Building Report" (http://hsewsf.sedsh.gov.uk/hslive/hsstart?P_HBNUM=28022). Historic Scotland. . Retrieved 2010-07-06.

[56] "Ravelston Garden, nos. 1-48: Listed Building Report" (http://hsewsf.sedsh.gov.uk/hslive/hsstart?P_HBNUM=30264). Historic Scotland. . Retrieved 2010-07-06.

[57] "Harbour House: Listed Building Report" (http://hsewsf.sedsh.gov.uk/hslive/hsstart?P_HBNUM=45507). Historic Scotland. . Retrieved 2010-07-06.

[58] "National War Museum of Scotland: Listed Building Report" (http://hsewsf.sedsh.gov.uk/hslive/hsstart?P_HBNUM=48223). Historic Scotland. . Retrieved 2010-07-06.

[59] "Ostaig Farm Square known as Sabhal Mor Ostaig: Listed Building Report" (http://hsewsf.sedsh.gov.uk/hslive/ hsstart?P_HBNUM=13985). Historic Scotland. . Retrieved 2010-07-06.

[60] "Cathedral of St John The Divine (Episcopal): Listed Building Report" (http://hsewsf.sedsh.gov.uk/hslive/hsstart?P_HBNUM=38849). Historic Scotland. . Retrieved 2010-07-06.

[61] "The Belmont Picturehouse: Listed Building Report" (http://hsewsf.sedsh.gov.uk/hslive/hsstart?P_HBNUM=20132). Historic Scotland. . Retrieved 2010-07-06.

[62] "Craigend Castle: Listed Building Report" (http://hsewsf.sedsh.gov.uk/hslive/hsstart?P_HBNUM=50821). Historic Scotland. . Retrieved 2010-07-06.

[63] The Historical Association. "The National Heritage List for England has gone live" (http://www.history.org.uk/resources/ public_news_1121.html). . Retrieved 23 May 2011.

[64] "The National Heritage List for England" (http://list.english-heritage.org.uk/). English Heritage. . Retrieved 23 May 2011.

[65] "Historic and Listed Buildings" (http://www.historic-scotland.gov.uk/index/heritage/historicandlistedbuildings.htm). Historic Scotland. . Retrieved 7 June 2011.

[66] http://jura.rcahms.gov.uk/PASTMAP/start.jsp

[67] http://www.britishlistedbuildings.co.uk/

[68] "Northern Ireland Buildings Database" (http://www.doeni.gov.uk/niea/built-home/recording/historic_buildings_r/buildings_database. htm). Northern Ireland Environment Agency. . Retrieved 7 June 2011.

[69] "Images of England FAQs" (http://www.imagesofengland.org.uk/Faqs/default.aspx). *English Heritage*. . Retrieved 8 June 2011.

[70] "Buildings at Risk" (http://www.english-heritage.org.uk/caring/heritage-at-risk/buildings-at-risk/). *English Heritage*. . Retrieved 24 May 2011.

[71] "What is Heritage at Risk?" (http://www.helm.org.uk/server/show/nav.19627). *Helm*. . Retrieved 8 June 2011.

[72] "Heritage at Risk" (http://risk.english-heritage.org.uk/2010.aspx). . Retrieved 24 May 2011.

[73] "Buildings at Risk" (http://www.buildingsatrisk.org.uk/BAR/). *Royal Commission on the Ancient and Historical Monuments of Scotland*. . Retrieved 8 June 2011.

[74] "Tackling Wales' buildings at risk" (http://www.cadw.wales.gov.uk/default.asp?id=21&NewsId=284). *Cadw*. 1 Novemebr 2009. .
      Retrieved 8 June 2011.

External links

- English Heritage on listed buildings (http://www.english-heritage.org.uk/caring/listing/listed-buildings/)
- National Heritage List for England: map-based database of listed buildings, scheduled monuments etc in England
  (http://list.english-heritage.org.uk/)
- PASTMAP Map-based database of listed buildings, scheduled monuments etc in Scotland (http://jura.rcahms.
  gov.uk/PASTMAP/start.jsp)
- Images of England:photographs of listed buildings (http://www.imagesofengland.org.uk/)
- Cadw, Wales (http://cadw.wales.gov.uk/splash?orig=/)

Churston_Ferrers

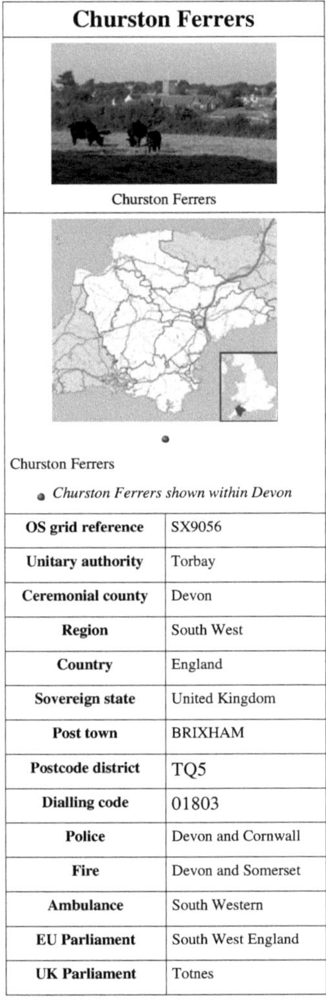

Churston Ferrers	
Churston Ferrers	
Churston Ferrers	
Churston Ferrers shown within Devon	
OS grid reference	SX9056
Unitary authority	Torbay
Ceremonial county	Devon
Region	South West
Country	England
Sovereign state	United Kingdom
Post town	BRIXHAM
Postcode district	TQ5
Dialling code	01803
Police	Devon and Cornwall
Fire	Devon and Somerset
Ambulance	South Western
EU Parliament	South West England
UK Parliament	Totnes

Churston Ferrers is a historic civil parish[1] within Torbay, in Devon, England. It contains the two villages of *Churston*, a coastal village, and the now larger Galmpton. It is situated in between Paignton and Brixham.

Churston railway station is on the Paignton and Dartmouth Steam Railway. Churston Ferrers is also home to a very well respected grammar school, Churston Ferrers Grammar School.

The village golf club's president is the Welsh former snooker world number one, Ray Reardon. Churston Court Inn dates to the 12th century and is a Grade I listed building.

Cultural references

In Agatha Christie's *The ABC Murders*, the third murder takes place in Churston.

References

[1] http://www.devon.gov.uk/historicchurstonferrers

Dartmouth_Steam_Railway

Dartmouth Steam Railway	
Locale	Paignton, Devon, England
Terminus	Kingswear
Commercial operations	
Name	Kingswear branch
Built by	Dartmouth and Torbay Railway
Original gauge	7 ft 0 $\frac{1}{4}$ in (2140 mm)
Preserved operations	
Operated by	Dart Valley Railway
Stations	4
Length	6.5 miles (**unknown operator: u'strong'** km)
Preserved gauge	4 ft 8 $\frac{1}{2}$ in (1435 mm)
Commercial history	
Opened	1859
1864	Line completed
1892	Converted to standard gauge
Closed	1972
Preservation history	
1972	Sold to Dart Valley Railway
Website	
[1]	

The **Dartmouth Steam Railway**, formerly known as the **Paignton & Dartmouth Steam Railway** is a 6.5 miles (**unknown operator: u'strong'** km) heritage railway on the former Kingswear branch line between Paignton and Kingswear in Torbay, Devon, England.

Due to the location of this line – at the heart of the English Riviera – much of the railway's business is summer tourists from the resorts of Torbay who are transported to Kingswear railway station from where a ferry takes them across the River Dart to the historic town of Dartmouth.

The line is owned and operated by Dart Valley Railway plc, who also own Dart Pleasure Craft Limited. Dart Pleasure Craft, who also trade as *River Link*, operates the Dartmouth Passenger Ferry between Kingswear and Dartmouth, together with river and coastal cruises from Dartmouth, many of which connect with the railway.[2] [3] As such, it is unusual amongst preserved railways in that it is a for-profit operation, and does not rely on volunteer labour or charitable donations.

History

Kingswear branch

The line was built by the Dartmouth and Torbay Railway, opening to Brixham Road station on 14 March 1861 and on to Kingswear on 10 August 1864.[4] The Dartmouth and Torbay Railway was always operated by the South Devon Railway and was amalgamated with it on 1 January 1872. This was only short-lived as the South Devon Railway was in turn amalgamated into the Great Western Railway on 1 February 1876. Brixham Road became a junction and was renamed "Churston" on 1 January 1868 when the independent Torbay and Brixham Railway opened its short line.[5]

The line was single-track except for a crossing loop at Churston. It had been built using the 7 ft 0 $\frac{1}{4}$ in (2140 mm) broad gauge, but on 21 May 1892 was converted to 4 ft 8 $\frac{1}{2}$ in (1435 mm) standard gauge.

West of Greenway Tunnel the railway was originally carried across two creeks on low timber viaducts, that at Longwood being 200 yards (**unknown operator: u'strong'** m) long and Noss being 170 yards (**unknown operator: u'strong'** m). These were demolished after the line was moved inland around the creeks on 20 May 1923.

A station was opened at Goodrington Sands, south of Paignton, on 9 July 1928. A second new halt was constructed at Broadsands Halt at the same time but was never opened for timetabled trains.[6] Park Sidings opened alongside Paignton Station in 1930 to give more room to stable carriages. A goods depot opened south of the station the following year, and the running line was doubled as far as Goodrington Sands.[4]

The Great Western Railway was nationalised into British Railways on 1 January 1948. Further carriage sidings to handle the heavy traffic on summer Saturdays were opened at Goodrington in 1956 and a turntable installed there in the following year.

Except for peak season trains, most services from 18 April 1966 operated as a shuttle service from Paignton; Sunday trains were withdrawn from 24 September 1967, although some were run during the summer of the following year. The Brixham branch closed on 13 May 1963 and the crossing loop at Churston was closed on 20 October 1968.[6]

Heritage railway

In 1968 it was formally proposed to the Ministry of Transport that the line from Paignton should be closed entirely but instead, on 30 December 1972, the line was sold to the Dart Valley Railway company, which at that time operated the nearby heritage railway that subsequently became the South Devon Railway. A winter service was operated from 1 January 1973 but from the end of that summer it became a purely seasonal operation. The purchase price of the railway was £250,000 and a further £25,000 was paid for signalling alterations at Paignton. Most of this was recouped from the sale of The Royal Dart Hotel at Kingswear and other surplus land.

An independent station alongside the main station at Paignton, known as "Queens Park", was opened to serve the Kingswear trains on the site of the old Park Sidings. The line was marketed at the time as the "Torbay Steam Railway",[4] but this has since been changed to "Paignton & Dartmouth Steam Railway". It remains the property of The Dart Valley Light Railway Company plc.

A loop was reinstated at Churston in 1979 using colour-light signals, and in 1981 the turntable from the British Rail sidings at Goodrington was moved there. In 1991 the control of all signalling was moved to a new panel at Britannia Crossing near Kingswear. A locomotive workshop was opened at Churston in 1993 and a carriage shop opened there three years later.

In 2007 the second track and carriage sidings were reinstated at Goodrington Sands to give more space for storing rolling stock.

The Association of Train Operating Companies included Brixham one of fourteen towns that, based on 2009 data, would benefit from a new railway service. This would be an extension of the First Great Western service on the Riviera Line from Exmouth as far as Churston, which would then act as a railhead for Brixham. It would also serve other housing developments in the area since the opening of the steam railway, and may require the doubling of that line between Paignton and Goodrington Sands.[7]

Operation

The operational base is at Paignton, where an engine shed is part of the station buildings. Heavy overhauls are undertaken at Churston where there is a locomotive workshop on the west side of the line, and a carriage shop and turntable on the east side.

Signalling is by electric multiple-aspect signals controlled from a panel at Britannia Crossing. The level crossing at that site is supervised by the signalman at the panel, but that at Sands Road, just outside Paignton station, is operated locally by the train crew.

All stations have booking offices but those at Goodrington Sands and Churston are only open at busy times and tickets are issued on the train most of the time.

The steam railway also operates a small fleet of buses, many of which are open-top, on routes in connection with the railway and ferry.

Route

The route is described facing forwards from Paignton to Kingswear, which puts the sea on the left and the River Dart on the right.

King Edward I at Saltern Cove

Cutting at the approach to Greenway tunnel

The line starts from its own platform at Paignton. The shed for operational locomotives is built into the south end of the station building, although coaling is done at the north end alongside the entrance used by passengers. Immediately beyond the station the line crosses Sands Road on a level crossing. The second track, on the right, is used by Network Rail to access their carriage sidings. There is a crossover between the two lines that allows trains from Network Rail to run through onto the steam railway.

Opposite the Network Rail carriage sidings on the right is a siding used by the steam railway to store engineering equipment. The train now calls at Goodrington Sands station, behind the platform to the right are more sidings which were transferred to the steam railway in 2007. The second platform is not currently in use, all trains calling at the one on the left.

Beyond the station the line starts its climb up a steep gradient behind the beach huts that line Goodrington Beach. The 630 miles (**unknown operator: u'strong'** km) South West Coast Path follows alongside the line on the right. After a small headland the train passes the secluded Saltern Cove and Armchair Rock, then swings right to pass over first 72 yards (**unknown operator: u'strong'** m) Broadsands Viaduct and then swings inland over the 148 yards (**unknown operator: u'strong'** m) Hookhills Viaduct before reaching the line's summit at Churston. On the approach to the station the turntable is seen on the left; this is where the Brixham branch line used to join the Kingswear branch.

Approaching Greenway Tunnel

From here the line drops down through Greenway Tunnel, beyond which the River Dart appears on the right. Once down to nearly river level it passes over Britannia Crossing, a level crossing over the A379 road on its approaches to the Dartmouth Higher Ferry. It is from the signal box here that the signalling for the whole line is controlled. Shortly after Britannia Crossing, the line arrives at Kingswear station. The far end of the platform is covered by a wooden train shed in the style favoured by Isambard Kingdom Brunel, although he died more than four years before the station was built. The boat- and car-park alongside the station was once a busy rail-served quayside goods yard.

The ferry across the Dart to Dartmouth leaves from the slipway which is behind the hotel next to the station. Dartmouth railway station is unique in that it has never seen a train as passengers have always arrived at the station by means of the ferry from Kingswear.

Rolling stock

Operational locomotives are shown in **bold text**. The status of locomotives may change at short notice and so this list may not represent the current position. Only "Lydham Manor" carried a name before preservation, but since being preserved on the railway all the others have also been given names from classical mythology. The liveries carried are generally based on, but not the same as, those carried before preservation.

Steam locomotives

- **4277 "Hercules"**: A GWR 4200 Class 2-8-0T, painted in GWR livery. Its boiler certificate expires in 2018
- 4555 "Warrior": A GWR 4500 Class 2-6-2T, painted in Great Western Railway (GWR) green livery. It is currently out-of-service and awaiting overhaul.
- 4588 "Trojan": A GWR 4575 Class 2-6-2T, a later version of the 4500 Class with larger tanks. It is painted in GWR green livery but is currently out of service awaiting an overhaul and is available for sale.
- **5239 "Goliath"**: A GWR 5205 Class 2-8-0T, painted in GWR livery. Its boiler certificate expires in 2017.
- **7827 "Lydham Manor"**: A GWR 7800 Class 4-6-0 locomotive painted in BR lined black. It was painted like this for the heritage weekend during the late spring bank holiday. This is the only regular locomotive that actually carried its name before preservation. Its boiler certificate expires in 2015.
- 75014 "Braveheart": A BR standard class 4 4-6-0, painted in BR black livery. It is currently out of service and undergoing an overhaul which is expected to be finished by 2015. It will then replace 7827 "Lydham Manor" when its boiler certificate expires

7827 *"Lydham Manor"*
in BR black livery running as No,7800

7827 *"Lydham Manor"*
in GWR green livery

Diesel locomotives

- **D2192 "Titan"**

  A British Rail Class 03 0-6-0 shunter. It is painted in BR black livery except it carries Dart Rail insignia.

- **D3014 "Samson"**

  A larger British Rail Class 08 0-6-0 shunter. It carries BR green livery except for Dart Rail insignia. Used for shunting stock or engineering work.

- **D7535 "Mercury"**

D7535 *Mercury*

  A British Rail Class 25 Bo-Bo which is painted in BR green livery with Dart Rail insignia. Used on emergencies or special events.

Visiting locomotives

There are frequent visits during the summer season of Torbay Express excursions from Bristol Temple Meads. These are hauled right through to Kingswear by a variety of steam locomotives. In 2007 the regular locomotives were. *King Edward I* and *Tangmere* but in previous years *Union of South Africa* and *Bradley Manor* have also operated the service. Latterly, LMS Princess Royal Class 6201 Princess Elizabeth, GWR 4073 Class 5029 Nunney Castle and BR standard class 7 70013 Oliver Cromwell have visited the line and more recent visitors have included A1 Peppercorn Class Tornado 60163, A4 60019 Bittern and BR standard 71000 Duke Of Gloucester which hauled the final Torbay express excursion of 2010.

In 1973 and 1993 the famous *Flying Scotsman* hauled regular service trains on the line throughout the summer seasons.

Coaches

The railway has a fleet of 20 coaches with 19 available for service and a further coach has been refurbished to become a Brunel exhibit at Kingswear Station.

Eight of the coaches are former British Rail DMU Class 116 and class 117 trailer cars which give good views from their open saloons. Only seven of the DMU trailer cars are operational. The eighth coach was formerly in use as Santa's grotto but is now used as an Exhibition coach telling the story of the line, its building by Brunel and a brief history of the local area.

The Devon Belle Saloon

A Pullman observation saloon, originally built for the Devon Belle service, is used regularly on passenger services. It provides a unique view of the railway, although an additional charge is made to ride in it.

The rest of the Fleet (11) is made up of British Railways Mark 1 corridor coaches, 8 TSOs, 2 BSKs and a BSO. The livery of the carriages is a version of the former GWR "chocolate and cream" livery and most carry the name of either a female member of staff, or the name of a member of staff's child or grandchild. The brake carriages have had their former luggage vans converted to wheelchair accommodation.

References

[1] http://www.dartmouthrailriver.co.uk/

[2] "Paignton & Dartmouth Steam Railway" (http://www.paignton-steamrailway.co.uk/). Dart Valley Railway plc. . Retrieved 2008-09-18.

[3] "River Link - Devon's River Dart Cruises" (http://www.riverlink.co.uk/). Dart Pleasure Craft Limited. . Retrieved 2008-09-18.

[4] Potts, C R (1998). *The Newton Abbot to Kingswear Railway (1844 - 1988)*. Oxford: Oakwood Press. ISBN 0-853613-87-7.

[5] Potts, CR (2000). *The Brixham Branch*. Usk: Oakwood Press. ISBN 0-853615-56-X.

[6] Oakley, Mike (2007). *Devon Railway Stations*. Wimbourne: The Dovecote Press. ISBN 978-1-904349-55-6.

[7] "Connecting Communities - expanding access to the rail network" (http://www.atoc.org/general/ConnectingCommunitiesReport_S10.
     pdf). Association of Train Operating Companies. 2009. . Retrieved 2009-09-16.

External links

- Dartmouth Steam Railway and River Boat Company website (http://www.dartmouthrailriver.co.uk/)

Churston_railway_station

<table>
<tr><td colspan="2" align="center">Churston</td></tr>
<tr><td colspan="2" align="center"></td></tr>
<tr><td colspan="2" align="center">Location</td></tr>
<tr><td>Place</td><td>Churston Ferrers, Devon</td></tr>
<tr><td>Area</td><td>Torbay</td></tr>
<tr><td>Coordinates</td><td>50°23′46″N 3°33′24″W</td></tr>
<tr><td>Grid reference</td><td>SX894562</td></tr>
<tr><td colspan="2" align="center">Operations</td></tr>
<tr><td>Original company</td><td>Dartmouth and Torbay Railway</td></tr>
<tr><td>Managed by</td><td>South Devon Railway</td></tr>
<tr><td>Pre-grouping</td><td>Great Western Railway</td></tr>
<tr><td>Post-grouping</td><td>Great Western Railway</td></tr>
<tr><td>Operated by</td><td>Paignton and Dartmouth Steam Railway</td></tr>
<tr><td>Platforms</td><td>2</td></tr>
<tr><td colspan="2" align="center">History</td></tr>
<tr><td>1861</td><td>Opened (as Brixham Road)</td></tr>
<tr><td>1868</td><td>Brixham branch opened</td></tr>
<tr><td>1972</td><td>Preserved</td></tr>
<tr><td colspan="2" align="center">Stations on heritage railways in the United Kingdom</td></tr>
<tr><td colspan="2" align="center">A B C D E F G H I J K L M N O P Q R S T U V W X Y Z</td></tr>
</table>

Churston railway station is on the Dartmouth Steam Railway, a heritage railway in Torbay, Devon, England. It is on the main road to Brixham and close to the villages of Galmpton and Churston Ferrers.

History

Auto-train for Brixham in 1958

Before preservation

The Dartmouth and Torbay Railway from Paignton railway station to Churston was opened for passengers on 14 March 1861 and for goods traffic on 1 April 1861. The station was known as **Brixham Road** at the time, and the line was extended to Kingswear railway station on 16 August 1864. The initial single platform was supplemented by a second in 1865.[1] The Dartmouth and Torbay Railway was always operated by the South Devon Railway Company and was amalgamated with it on 1 January 1872. This was only short lived as it was in turn amalgamated into the Great Western Railway on 1 February 1876.

In the meantime the station had changed its name to Churston when an independent branch line had been opened by the Torbay and Brixham Railway to serve the latter town on 28 February 1868 .[2] The station was now a junction but the goods shed had to be moved to a new site alongside the Brixham line to make room for a short bay platform to accommodate the Brixham trains. Sidings were added to allow for the goods traffic handled on the branch, including a busy trade in fish. The Torbay & Brixham Railway was taken over by the Great Western Railway on 1 January 1883.

The lines had been built using the 7 ft 0 $\frac{1}{4}$ in (2140 mm) broad gauge, but on 21 May 1892 were closed for the weekend to be converted to 4 ft 8 $\frac{1}{2}$ in (1435 mm) standard gauge. The following year saw the platforms lengthened and a new signal box constructed. The platforms were further lengthened and a new signal box opened on 9 February 1913 to control the now extended crossing loop.

The Great Western Railway was nationalised into British Railways on 1 January 1948. The Brixham branch closed on 13 May 1963, but the Kingswear service continued but Sunday trains no longer called at Churston after the 1967 summer season. General freight traffic was withdrawn on 14 June 1965 although coal was still handled until 4 December 1967. The crossing loop was taken out of use on 20 October 1968 and the signal box closed.

In preservation

The line was sold to the Dart Valley Light Railway plc on 30 December 1972, which operated another nearby railway at Buckfastleigh. Since then Churston has become an important centre for engineering on the railway.

The signal box was reopened in 1979 to control a new crossing loop, albeit with electric multiple-aspect signals, and the following year the former Brixham bay platform was relaid. In 1981 the turntable from Goodrington was moved to a position alongside the Brixham junction at Churston. The signal box was closed again in 1991[3] when control of the whole line was transferred to Britannia Crossing at Kingswear. A locomotive workshop was built behind the Up (towards Paignton) platform in 1993 and the station building restored and reopened. The Brixham bay platform was then covered by a carriage workshop in 1996.

Plans

The Association of Train Operating Companies included Brixham one of fourteen towns that, based on 2009 data, would benefit from a new railway service. This would be an extension of the First Great Western service over the Riviera Line from Exmouth as far as Churston, which would then act as a railhead for Brixham. It would also serve other housing developments in the area since the opening of the steam railway, and may require the doubling of that line between Paignton and Goodrington Sands.[4]

Description

The main entrance is onto the platform mainly used by trains towards Kingswear. Signalling allows trains to use this in both directions but in practice trains towards Paignton depart from the opposite platform, which is reached by a footbridge. The station buildings are built of solid masonry with a large canopy integral with the roof. The local paper described them on the opening day "as unarchitectural as any Goth could wish".[5]

Alongside the main platform, at the Paignton end, a modern workshop stands on the site of the platform once used by Brixham trains; this is used for carriage repairs. On the same side, beyond the modern bridge carrying the road to Brixham, are some sidings and the turntable. Opposite the carriage workshop, behind the Paignton platform, is another workshop which is used for heavy repairs to locomotives.

The former Railway Hotel at the end of the station approach road is now a 4-star guest house.[6]

Cultural references

Churston station was the location for the opening scene in *The System*, a 1964 film. Several of the main characters arrive at the last minute and jump onto a Diesel Multiple Unit that is just pulling out towards Kingswear. After the opening credits, they then get off a different train at Brixham. Agatha Christie lived nearby, and set some stories in and around the area, Churston is the 'C' in the The A.B.C. Murders.

Services

A seasonal service of steam hauled trains operates between Paignton and Kingswear.

Preceding station	🚂 Heritage railways	Following station
Goodrington Sands	Dartmouth Steam Railway	Kingswear

References

Further reading

- Beck, Keith; Copsey, John (1990). *The Great Western in South Devon*. Didcot: Wild Swan Publication. ISBN 0-906867-90-8.
- Cooke, RA (1984). *Track Layout Diagrams of the GWR and BR WR, Section 14: South Devon*. Harwell: RA Cooke.

This station offers access to the South West Coast Path	
Distance to path	
Next station anticlockwise	Goodrington Sands 3 miles (**unknown operator: u'strong'** km)
Next station clockwise	Kingswear 12 miles (**unknown operator: u'strong'** km)

Article Sources and Contributors

Galmpton,_Torbay *Source*: http://en.wikipedia.org/w/index.php?title=Galmpton%2C_Torbay *Contributors*: Finavon, Herbythyme, MatGB, Pebleco, Snigbrook, 8 anonymous edits

Torbay *Source*: http://en.wikipedia.org/w/index.php?title=Torbay *Contributors*: Acalamari, Adziura, Amakuru, Andrew Dalby, Andrewjlockley, Anwar saadat, Apoc2400, Arpingstone, Beautytime, Bleaney, BlueMoses, Bob Re-bom, Bsrboy, Chapmlg, Chrislintott, Chzz, Closedmouth, Cnyborg, CottrellS, Craigy144, Danjamz, Dawd, Deflective, Dpaajones, DéRahier, Edward, Esperant, Espresso Addict, Funnyhat, Geof Sheppard, Graham87, Grstain, Grutness, Herbythyme, Hesperian, Highwayville, Icairns, Jmorrison230582, Jolly Janner, Jordo72, Kane5187, Keith D, Keith Edkins, Kwamikagami, Lambdoid, Lemonade100, Lewisskinner, Lfh, LindaHill, MRSC, Magioladitis, Marketingcubelab, Marknew, Marky-Son, Matt Adore, MattBlay, Mav, Miller17CU94, Modal Jig, Morwen, MountainRanger74, MuZemike, NigelR, Nightrider 83, Nilfanion, Nzpcmad, Old Moonraker, Oliver Chettle, Orioane, Paigntonuk, Paul.m.hutchinson, Paulleake, Pcpcpc, Penhulahoop, Pharaoh of the Wizards, Philcrbk, Philip Trueman, Plrk, Prawnz, Randalthor1812, Redf0x, Rich Farmbrough, SP-KP, Safalra, Sceptic, Simple Bob, Skinsmoke, Smalljim, Snigbrook, Sophie means wisdom, SqueakBox, Starbois, Steinsky, Stephen Burnett, Stephenb, Tearlach, Template namespace initialisation script, The Singing Badger, The flying pasty, Torquayboy54, Tqywalker, Tuesdaily, Vafthrudnir, Warofdreams, Wereon, Wideperspective, Winston365, Zzuuzz, 81 anonymous edits

Devon *Source*: http://en.wikipedia.org/w/index.php?title=Devon *Contributors*: 10.0.0.60, 217.135.230.12, 212.248.133.xxx, 80.255, A E Francis, ACEOREVIVED, Aaron north, Abfabmedia, Abisharan, Acalamari, Achangeisasgoodasa, Addshore, Adsomvilay, Aetheling1125, Ahoerstemeier, Airunp, Aitias, Akamad, Akhilleus, Alansohn, Alex1011, Amazonien, Amberroom, Andrea105, Angr, AnonMoos, Anwar saadat, AnyFile, Arpingstone, Artinyourarea, Aruton, AssociateAffiliate, Asterion, Auldlangsyne2010, Avoided, AxG, B3nz0, BB98uk, Bainseyy, Balin42632003, Balloonguy, Bazonka, Bcorr, Benjaminevans82, Bennelliott, Bethling, Big Bird, Bill1001, Binksternet, BlkBeard, Bob Re-born, Bobblehead, Boblybo, Bobo192, Boingboy, Bongwarrior, Booyabazooka, Bornintheguz, Boston, Bozoid, Bretagne 44, Bretonbanquet, BrianGV, Bryan Derksen, Bsrboy, Bwithh, Caponer, Carabinieri, Carby0, CarolGray, Casliber, Caulde, Century0, Ceyockey, Chanheigeorge, Charles Matthews, Cheesy mike, Cherry Tree Lane, Choccheesecake, Choess, Chris j wood, Christmas-jones, Chrizkerr2, Ciaccona, Citterio, Cjthellama, Clootie, Cnyborg, Conversion script, Conzie09, Coolhawks88, Coreyatschool, Craitman17, Cyndilauper, Cyr, Cyrius, D6, Dahliarose, Dainamo, Dana boomer, Dartelaar, Dartmoorpete, Davewild, David Biddulph, DavidFarmbrough, Davyth, Dawd, Ddstretch, Debby lacy, Debresser, Deepak D'Souza, Definition12, Deflective, DennyColt, DerHexer, Deunansek, Devahn58, Devonboy69, Devonian13, Devonplp, Devonsid, Dewnans, Diderot, DinosaursLoveExistence, Discospinster, Dismas, Divercol, Djr xi, Doops, Dougofborg, Dpaajones, Dpfwiki, Drake 1992, Drumguy8800, DuncanHill, Dusimpson, Dweller, Dynamo-dmw, DynamoDegsy, DéRahier, Eboracum, Edderso, Ehrenkater, Elassint, Elf, Elinorlol, Ellywa, Emuchick, Enchanter, EoGuy, Epbr123, Ericamick, Eugene-elgato, Evercat, Evil saltine, Exeterhockey, Fastily, Favonian, Felix Folio Secundus, Fieldday-sunday, Flatterworld, Foggy dew, Fraggle81, Francs2000, Fremsley, Fruitbat101, GLaDOS, GR Davies, GSTQ, Gabbe, GainLine, Gaius Cornelius, Gaius Octavius Princeps, Galoubet, GazMan7, Geanixx, GeoGreg, Geof Sheppard, GeorgeWClinton, Gerble1000, Gerren, Ghaly, Ghmyrtle, Giss, Glasgowbhoy69, Gogo Dodo, Gomm, GordyB, Goto, GraemeLeggett, Graham87, Greatestrowerever, Greentubing, Grstain, Gtstricky, Gwernol, Habj, Hadal, HappyCamper, HappyFace, Hatherleighholidays, Henryjimdix, Herbythyme, Heron, Historyfool, Hitman504, Hitssquad, Hmains, HooperBandP, Hotlorp, Howard Alexander, Howcheng, Hushpuckena, Hydrogen Iodide, Idiot10, Ilikeeatingwaffles, InfinityAndBeyond, Iridescent, J.delanoy, JS500, Jackhynes, Jaguar, JamaicanBacon1, JamesFitz, Jao, Jargon777, Jcmurphy, Jdw817, Jess.goddard, Jguk 2, Jimp, Jmorrison230582, Johnluisocasio, Jolly Janner, JonC, Jonny pop, Joowwww, Joseph Solis in Australia, Jreferee, JzG, Kaihsu, Kanguole, Katherine Shaw, Keith Edkins, Kelisi, Kernowabc, Kipperfield, KirkEN, Kisburgh, KnightRider, Knowinitall21, Korovioff, Kotla Mohsin Khan, Krang, Krish387, Kummi, Kwamikagami, LeaveSleaves, Leonard^Bloom, Lethesl, Lethetruthbeknown, Levihashman, Lightmouse, Lloydpick, Lost tourist, Lozleader, Lynbarn, MER-C, MHenbert, MJCdetroit, MPF, MRSC, MZMcBride, MacRusgail, Magnus, Magnus Manske, Mais oui!, MarcoTolo, Marek69, Marj Tiefert, Marknew, Marnanel, Matdrodes, Materialscientist, Mattbr, Mattcridland, Mattcridland1, Mattis, Mav, Mayumashu, Mbz1, Meaty Weenies, Mehmet12, Mel Etitis, Mentifisto, Mervyn, Mhockey, Mic, Michael Glass, Michael Hardy, Midgley, Mike s, Mikenorton, Mild Bill Hiccup, Mlk, Mlm42, Modal Jig, Moncrief, Morwen, Mr Stephen, Mr pand, MrTom, Mystery1229, Natl1, Naveira, NawlinWiki, Nenniu, Netalarm, Netoholic, Neurolysis, Nevilley, Nick xylas, NigelR, Nilfanion, Nonstoptraveller, Norm, Northamerica1000, NotAnonymous0, Oaken, Ohconfucius, Ohnoitsjamie, OldSpot61, Orioane, OshareKei, Our Phellap, Owain, Owain.davies, OwenBlacker, PDMB, PIrish, Pablothegreat85, Palica, Pdiddyjr, Pedant17, Pharillon, Philip Trueman, PhoenixnumbaOne, Picaroon, PierreAbbat, Pigman, PirateSmackK, Plucas58, Plutonium27, PoPFanboy, Poiuytre, Pollinator, Polylerus, Popstock, Postlebury, Profjack, PseudoOne, PseudoSudo, Pwqn, QuartierLatin1968, RB972, RabidWolf, RadicalBender, Rammsteinlieber, RandomXYZb, Rawrrawrdinosaurlawl, Renata, Rettetast, Revth, RewinderPaul, RexNL, RichardPhillipsb38, Richardk74, Richardmayne, Richardsalter, Rick@lcl.nl, Riwnodennyk, Rjakw, Rjmdev, Rjwilmsi, Robertvan1, Rodw, Ronhjones, Rrburke, RupertB, S.Örvarr.S, SAMbo, SE7, SELIbydate, SP-KP, Saga City, Saintswithin, Sakurambo, Salljf77, Samg96, Sampo Tiensuu, Sardonicone, Saywhat07, Sceptre, SchuminWeb, Seb26, Seglea, Shailendrapalsinghrathore, Shanes, Showjumpersam, Silversam, Simon1959, Simple Bob, Sionus, Siswrn, Sjc, Sjorford, Skier Dude, Skrofler, Skybunny, Small is beautiful, Smalljim, Snappy, Snowmanradio, Soccerrocker1996, Soloist, Someone1235679, Sophie means wisdom, Soyuz113, Spolky, SpookyMulder, Steinsky, Stephenb, Stevebritgimp, Steven14rule, Stewartadcock, Stringops, Stumpypost, Sumitha mlh, Sunray, Surfgatinho, Surlyduff50, Svitapeneela, Swithlander, T0m, Tayloradolphe, Tbone762, Template namespace initialisation script, Tertius0, TexMurphy, TexasAndroid, The Geography Elite, The Master of Mayhem, The Rambling Man, The Thing That Should Not Be, Thenudist, Thomas Blomberg, Thomas Larsen, Timrollpickering, Timwi, Tinmuff, Tobias Hoevekamp, Tomclarke, Tomgreeny, Tomsega, Tony in Devon, Trebor, Trident13, Trilobite, Tripod86, Ttony21, Tungol, Twantooq, Twistlethrop, Tyroshi 127, Uncle Dick, Undertheheavens, Unyoyega, VampWillow, VanillaBear23, Vicebdon, Vranak, Vuzzkite, WHATaintNOcountryIeverHEARDofDOtheySPEAKenglishINwhat, WOSlinker, Warofdreams, Wayne Slam, Wclark, Wcwtlc94, Wereon, Wetman, Wh0834, WhisperToMe, WikHead, Wiki Raja, WikiBern, Wikipelli, Willking1979, Woohookitty, Wsupermain2, Xtc2000, Yesideez, ZPM, Zangar, Zburh, Zoicon5, Τασουλα, 740 anonymous edits

Brixham *Source*: http://en.wikipedia.org/w/index.php?title=Brixham *Contributors*: Acalamari, Achangeisasgoodasa, Andrew Dalby, Andycjp, Angela, Arpingstone, Bevo74, BritishWatcher, Britmax, Brixhamtc, Bsrboy, CMat, CanOfWorms, Canglesea, CarolGray, Caulde, Chris the speller, CommonsDelinker, Craitman17, Dartnick, DaveSmedley, Dawd, Deflective, Delberto, Djencore, Djencpulse, Dublineramerican, Dynamo-dmw, Embrixde, Epipelagic, Geof Sheppard, Geoking66, Gillean666, Happysailor, Hephaestos, Herbythyme, I LIKE AMY POTGIETER IN BED, Iohannes Animosus, Iridescent, Iwool, Jimmy Pitt, John, Jolly Janner, Jonjoe, Joseph Solis in Australia, JustAGal, JzG, Jza84, Keith Edkins, Kidd Arda, Kusunose, Kwamikagami, Kwiki, LarryJeff, LindaHill, Lozleader, Lupin, Magnus Manske, Mais oui!, Marketingcubelab, Mav, MountainRanger74, Mr Stephen, NigelR, Norm, Nv8200p, Oneblackline, Orioane, Paul-b4, Philhhaha, Pit-yacker, Rebelfilms, Redf0x, Rex Germanus, Rjwilmsi, SE7, SP-KP, SQL, Safalra, Sannse, Secretarmy, SimonP, Simple Bob, Skinsmoke, Sliggy, Smalljim, Someguy1221, Sophie means wisdom, SpaceFlight89, Stephen Burnett, THEN WHO WAS PHONE?, TempyIncursion, Thumperward, Trident13, Warofdreams, Webber91, Welsh, Wereon, William Avery, Windharp, Worcsworks, Yvwv, Ælfgar, 164 anonymous edits

Greenway_Estate *Source*: http://en.wikipedia.org/w/index.php?title=Greenway_Estate *Contributors*: Aimsworthy, Alisonjenenr, Cambyses, DrSativa, Eric-Wester, Esowteric, Felix Folio Secundus, Gareth E Kegg, Grstain, Grutness, Hailey C. Shannon, Hanbrook, Hebrides, Hugo999, Hyacinth Bucket, Jameschristopher, Jllm06, MilborneOne, Mmdolbow, MortimerCat, Oliver Chettle, Pcpcpc, Rossrs, Ser Amantio di Nicolao, Smalljim, Snigbrook, Tim!, Vernon39, Wayward, Wescbell, 5 anonymous edits

Listed_building *Source*: http://en.wikipedia.org/w/index.php?title=Listed_building *Contributors*: Aidan Croft, Alanmaher, Andreasegde, ArnoldReinhold, Art LaPella, B.dyck, Bankhallbretherton, Barliner, Barnabypage, Bbacambridge, Ben Ben, Ben davison, Bhoeble, BigJohnD, Bigjimr, Biscuittin, Bobadillaman, Bornhard74, Buckygooner, CalJW, Canol, Celtus, Ceyockey, Cezarika1, ChrisGualtieri, Chrisminter, Cindamuse, Closedmouth, Cnyborg, CommonsDelinker, Computerjoe, Conscious, Courcelles, Crisco 1492, Crouch, Swale, Daniel Newman, DarTar, David Edgar, DavidAnstiss, Daviessimo, Daytona2, Dhartung, Disambiguator, DisillusionedBitterAndKnackered, DonJay, Doncram, Dormskirk, DrBob, Dricherby, Drumhollistan, Dthomsen8, East718, Elkman, Emeraude, EoGuy, Ephebi, Erebus555, Erianna, Erik Kennedy, ErikTheBikeMan, Espresso Addict, Evil Eye, FGLawson, Forgotten the crackers, FrFintonStack, Fridaynight, Gaius Cornelius, George Burgess, Ghughesarch, Gjbarclay, Gnevin, Goldenlane, Grstain, Guinness2702, Guy Hatton, Handsaw, Harryboyles, Herbythyme, Honbicot, IdreamofJeanie, Ilikeeatingwaffles, Indisciplined, Irate, Iridescent, Israelboy, Ivolocy, JBellis, JRawle, Jack1956, JackyR, Jdforrester, JillyFfoulkes, Jimd, Jmlk17, John Fader, Jonathan Oldenbuck, Joshurtree, Jtgerman, K1Bond007, Keith D, Ken Gallager, Kilnburn, Kingleonidas1, Kintak, Kjlewis, Klaus with K, Kuxu, LC Revelation, LeaveSleaves, Leithp, Leszek Jańczuk, Lewisskinner, Lidos, Lightmouse, Lilac Soul, Logan, Lugnuts, Lumos3, Lupo, MER-C, Magnius, Mais oui!, Majorly, Malleus Fatuorum, Man vyi, Marcika, Mark Wheaver, Martarius, Maryambeheshty, Mattis, Mauls, MaxVeers, Mcginnly, Michael Hardy, Middenface, Mild Bill Hiccup, Minden251, Mjroots, Mok9, Mon2s, Moonraker, Mooretwin, Morwen, Mr Stephen, Mr. High school student, Mwmonk, Neddyseagoon, NicholasNCE, Nohano, O Fenian, OKTerrific, Of7271, Olof nord, Oneblackline, Oosoom, PamD, Paul W, Paulleake, Pbl1998, Penfold, Peterlewis, PhantomReinhart, Phlytr, Pigsonthewing, Piledhigheranddeeper, Plastikspork, Pokeronskis, Pterre, Pyrotec, Qef, Quasihuman, Qwghlm, R'n'B, RHaworth, Ragemanchoo, Rangoon1, Ratarsed, Red van man, Rettetast, Richard Harvey, Richhoncho, Rjwilmsi, Robert EA Harvey, Rodw, Rosser1954, Roydosan, Rphistoricscotland, Rschon, Rupert Clayton, Rwendland, Sam Blacketer, Samshaft, Scottishcinemas, Scouseben, Secret Saturdays, Secretlondon, Simple Bob, Slow Riot, Smb1001, Snigbrook, Solipsist, Spandrels, Steinsky, Steverance, Stevouk, Straw Cat, Streapadair, SynEx, Teatreez, Terryrs, Tgm ravelston, The Advocate, The Anome, Theserialcomma, Thumperward, Timbn1, Timc, TomArra, TonyW, Trovatore, Tubs uk, VampWillow, Vegaswikian, Viva-Verdi, Vjam, Voceditenore, Voyager, WOSlinker, Warofdreams, Wavelength, Welshleprechaun, Wereon, Wiccasha, Wiki alf, Willdow, Worley-d, XLerate, Xn4, Yefi, YellowMonkey, Zephyrus67, Zigger, Zundark, Zywakem, 262 anonymous edits

Churston_Ferrers *Source*: http://en.wikipedia.org/w/index.php?title=Churston_Ferrers *Contributors*: Armbrust, Billinghurst, Cavrdg, Celtic guardian91, Cobaltbluetony, Craitman17, Dr. Blofeld, Marcus22, MatGB, Skinsmoke, Sophie means wisdom, Timrollpickering, 6 anonymous edits

Dartmouth_Steam_Railway *Source*: http://en.wikipedia.org/w/index.php?title=Dartmouth_Steam_Railway *Contributors*: 6024kingedward1, 7severn7, Alex1011, AmosWolfe, Biscuittin, Bjh21, Britmax, Broomhalla, Chase me ladies, I'm the Cavalry, Chris j wood, David Biddulph, DavidJones, DuchessofSutherland, Duncharris, EdJogg, Felix Folio Secundus, Fredrik, Geof Sheppard, Griffin700, Gwernol, HandigeHarry, HappyCamper, Herbythyme, I Do Care, JBTEvans, JHunterJ, Jafeluv, Kane5187, Kevin, Leonard^Bloom, Malcolm Morley, Marcus22, NE2, Ning-ning, Oliver Chettle, Our Phellap, Peter Horn, Peter Shearan, Pigman, Poltair, Postlebury, Redrose64, Rjwilmsi, Rkilpin, SilkTork, Slaffter, Starbois, Thryduulf, Tony May, Trident13, Twiceuponatime, Vanished user, Woohookitty, 131 anonymous edits

Churston_railway_station *Source*: http://en.wikipedia.org/w/index.php?title=Churston_railway_station *Contributors*: Adambro, Bjh21, Geof Sheppard, Ground Zero, Gwernol, Keith D, Kevin B12, Lamberhurst, Marcus22, MatGB, Njt1982, Redrose64, Rich Farmbrough, Shortfatlad, Skinsmoke, Snigbrook, Thelb4, 4 anonymous edits

Image Sources, Licenses and Contributors

file:Devon UK location map.svg *Source*: http://en.wikipedia.org/w/index.php?title=File:Devon_UK_location_map.svg *License*: unknown *Contributors*: User:Nilfanion

File:Red pog.svg *Source*: http://en.wikipedia.org/w/index.php?title=File:Red_pog.svg *License*: unknown *Contributors*: Anomie

File:Torbay UK locator map.svg *Source*: http://en.wikipedia.org/w/index.php?title=File:Torbay_UK_locator_map.svg *License*: unknown *Contributors*: User:Nilfanion

File:Loudspeaker.svg *Source*: http://en.wikipedia.org/w/index.php?title=File:Loudspeaker.svg *License*: unknown *Contributors*: Bayo, Gmaxwell, Husky, Iamunknown, Mirithing, Myself488, Nethac DIU, Omegatron, Rocket000, The Evil IP address, Wouterhagens, 18 anonymous edits

File:Torbay view.jpg *Source*: http://en.wikipedia.org/w/index.php?title=File:Torbay_view.jpg *License*: unknown *Contributors*: User:Kicior99

File:English Riviera Tours JTD395P.jpg *Source*: http://en.wikipedia.org/w/index.php?title=File:English_Riviera_Tours_JTD395P.jpg *License*: unknown *Contributors*: User:Geof Sheppard

File:Flag of Devon.svg *Source*: http://en.wikipedia.org/w/index.php?title=File:Flag_of_Devon.svg *License*: unknown *Contributors*: User:Greentubing

File:Devon UK locator map 2010.svg *Source*: http://en.wikipedia.org/w/index.php?title=File:Devon_UK_locator_map_2010.svg *License*: unknown *Contributors*: User:Nilfanion

File:Devon county council logo small.svg *Source*: http://en.wikipedia.org/w/index.php?title=File:Devon_county_council_logo_small.svg *License*: unknown *Contributors*: Jolly Janner, Sakurambo

File:Devon Ceremonial Numbered.png *Source*: http://en.wikipedia.org/w/index.php?title=File:Devon_Ceremonial_Numbered.png *License*: unknown *Contributors*: Habj, Herbythyme, Kanguole, Marknew, Smalljim

File:torquay.devon.750pix.jpg *Source*: http://en.wikipedia.org/w/index.php?title=File:Torquay.devon.750pix.jpg *License*: unknown *Contributors*: Ardfern, Ianmacm, Nk, Powerek38, 3 anonymous edits

File:Heath.jpg *Source*: http://en.wikipedia.org/w/index.php?title=File:Heath.jpg *License*: unknown *Contributors*: Original uploader was MPF at en.wikipedia. Later version(s) were uploaded by Ahuskay, Wendyy, Tayybrittain at en.wikipedia.

File:Ilfracombe.jpg *Source*: http://en.wikipedia.org/w/index.php?title=File:Ilfracombe.jpg *License*: unknown *Contributors*: Original uploader was Arpingstone at en.wikipedia

File:Devon fields stitch.jpg *Source*: http://en.wikipedia.org/w/index.php?title=File:Devon_fields_stitch.jpg *License*: unknown *Contributors*: User:Herbythyme

File:Magnify-clip.png *Source*: http://en.wikipedia.org/w/index.php?title=File:Magnify-clip.png *License*: unknown *Contributors*: User:Erasoft24

Image:Pnies5.jpg *Source*: http://en.wikipedia.org/w/index.php?title=File:Pnies5.jpg *License*: unknown *Contributors*: User:Wsupermain2

File:Exeter Cathedral 2923rw.jpg *Source*: http://en.wikipedia.org/w/index.php?title=File:Exeter_Cathedral_2923rw.jpg *License*: unknown *Contributors*: User:Rüdiger Wölk

File:devon.brixham.750pix.jpg *Source*: http://en.wikipedia.org/w/index.php?title=File:Devon.brixham.750pix.jpg *License*: unknown *Contributors*: Original uploader was Arpingstone at en.wikipedia

File:Devon arms.png *Source*: http://en.wikipedia.org/w/index.php?title=File:Devon_arms.png *License*: unknown *Contributors*: Alex1011, G.dallorto, Hogweard, Ilyaroz, Jolly Janner, Magul, Ssolbergj

File:westwardho.beach.arp.750pix.jpg *Source*: http://en.wikipedia.org/w/index.php?title=File:Westwardho.beach.arp.750pix.jpg *License*: unknown *Contributors*: Chris j wood, Man vyi, Marknew, Nilfanion, Thuresson

Image:brixham.devon.750pix.jpg *Source*: http://en.wikipedia.org/w/index.php?title=File:Brixham.devon.750pix.jpg *License*: unknown *Contributors*: Original uploader was Arpingstone at en.wikipedia

Image:BrixhamLookingWest750px.jpg *Source*: http://en.wikipedia.org/w/index.php?title=File:BrixhamLookingWest750px.jpg *License*: unknown *Contributors*: Original uploader was TempyIncursion at en.wikipedia

File:Breakwater light.jpg *Source*: http://en.wikipedia.org/w/index.php?title=File:Breakwater_light.jpg *License*: unknown *Contributors*: User:Herbythyme

File:Brixham.hind.750pix.jpg *Source*: http://en.wikipedia.org/w/index.php?title=File:Brixham.hind.750pix.jpg *License*: unknown *Contributors*: User:Akigka

File:Brixham railway station 1910213 26d7e678.jpg *Source*: http://en.wikipedia.org/w/index.php?title=File:Brixham_railway_station_1910213_26d7e678.jpg *License*: unknown *Contributors*: Ben Brooksbank

File:Greenway2010-245.jpg *Source*: http://en.wikipedia.org/w/index.php?title=File:Greenway2010-245.jpg *License*: unknown *Contributors*: User:MilborneOne

Image:Greenway House, Devon, England.jpg *Source*: http://en.wikipedia.org/w/index.php?title=File:Greenway_House,_Devon,_England.jpg *License*: unknown *Contributors*: David Hawgood

Image:DSCN2354GreenwayVictorianGreenhouse.jpg *Source*: http://en.wikipedia.org/w/index.php?title=File:DSCN2354GreenwayVictorianGreenhouse.jpg *License*: unknown *Contributors*: User:Vernon39

Image:DSCN2353Greenway.jpg *Source*: http://en.wikipedia.org/w/index.php?title=File:DSCN2353Greenway.jpg *License*: unknown *Contributors*: User:Vernon39

File:Vinery, Greenway House - geograph.org.uk - 191217.jpg *Source*: http://en.wikipedia.org/w/index.php?title=File:Vinery,_Greenway_House_-_geograph.org.uk_-_191217.jpg *License*: unknown *Contributors*: Nilfanion

Image:ForthRailwayBridge 27-06-2005 2150 TakenByEuchiasmus.JPG *Source*: http://en.wikipedia.org/w/index.php?title=File:ForthRailwayBridge_27-06-2005_2150_TakenByEuchiasmus.JPG *License*: unknown *Contributors*: Original uploader was Euchiasmus at en.wikipedia

Image:Buckingham Palace, London, England, 24Jan04.jpg *Source*: http://en.wikipedia.org/w/index.php?title=File:Buckingham_Palace,_London,_England,_24Jan04.jpg *License*: unknown *Contributors*: Ardfern, Bestiasonica, Conscious, Duffman, Hailey C. Shannon, Man vyi, Mtaylor848, Multichill, Voyager, 3 anonymous edits

File:Bank Hall Daffodils.jpeg *Source*: http://en.wikipedia.org/w/index.php?title=File:Bank_Hall_Daffodils.jpeg *License*: unknown *Contributors*: User:Bankhall

Image:Ccffc.jpg *Source*: http://en.wikipedia.org/w/index.php?title=File:Ccffc.jpg *License*: unknown *Contributors*: Original uploader was OKTerrific at en.wikipedia

File:MussendenTemple.jpg *Source*: http://en.wikipedia.org/w/index.php?title=File:MussendenTemple.jpg *License*: unknown *Contributors*: User:wuser10

Image:National Gallery of Scotland restitch1 2005-08-07.jpg *Source*: http://en.wikipedia.org/w/index.php?title=File:National_Gallery_of_Scotland_restitch1_2005-08-07.jpg *License*: unknown *Contributors*: Berrucomons, Jameslwoodward, JuTa, Klaus with K, Man vyi, Para, Tatata, Thierry Caro

Image:Wfm ibrox main stand.jpg *Source*: http://en.wikipedia.org/w/index.php?title=File:Wfm_ibrox_main_stand.jpg *License*: unknown *Contributors*: User:Finlay McWalter

File:Flag of France.svg *Source*: http://en.wikipedia.org/w/index.php?title=File:Flag_of_France.svg *License*: unknown *Contributors*: Anomie

File:Flag of Germany.svg *Source*: http://en.wikipedia.org/w/index.php?title=File:Flag_of_Germany.svg *License*: unknown *Contributors*: Anomie

File:Flag of Hong Kong.svg *Source*: http://en.wikipedia.org/w/index.php?title=File:Flag_of_Hong_Kong.svg *License*: unknown *Contributors*: Designed by

File:Flag of the Netherlands.svg *Source*: http://en.wikipedia.org/w/index.php?title=File:Flag_of_the_Netherlands.svg *License*: unknown *Contributors*: User:Zscout370

File:Flag of New Zealand.svg *Source*: http://en.wikipedia.org/w/index.php?title=File:Flag_of_New_Zealand.svg *License*: unknown *Contributors*: Adabow, Adambro, Arria Belli, Avenue, Bawolff, Bjankuloski06en, ButterStick, Denelson83, Donk, Duduziq, EugeneZelenko, Fred J, Fry1989, Hugh Jass, Ibagli, Jusjih, Klemen Kocjancic, Mamndassan, Mattes, Nightstallion, O, Peeperman, Poromiami, Reisio, Rfc1394, Shizhao, Tabasco, Transparent Blue, Väsk, Xufanc, Zscout370, 35 anonymous edits

File:Flag of Sweden.svg *Source*: http://en.wikipedia.org/w/index.php?title=File:Flag_of_Sweden.svg *License*: unknown *Contributors*: Anomie

File:Flag of the United States.svg *Source*: http://en.wikipedia.org/w/index.php?title=File:Flag_of_the_United_States.svg *License*: unknown *Contributors*: Anomie

Image:Churston church from the southeast - geograph.org.uk - 991078.jpg *Source*: http://en.wikipedia.org/w/index.php?title=File:Churston_church_from_the_southeast_-_geograph.org.uk_-_991078.jpg *License*: unknown *Contributors*: Derek Harper

Image:5239 near Goodrington.JPG *Source*: http://en.wikipedia.org/w/index.php?title=File:5239_near_Goodrington.JPG *License*: unknown *Contributors*: User:Geof Sheppard

Image:6024 Saltern Cove.jpg *Source*: http://en.wikipedia.org/w/index.php?title=File:6024_Saltern_Cove.jpg *License*: unknown *Contributors*: User:Geof Sheppard

File:Railway near Greenaway - geograph.org.uk - 23980.jpg *Source*: http://en.wikipedia.org/w/index.php?title=File:Railway_near_Greenaway_-_geograph.org.uk_-_23980.jpg *License*: unknown *Contributors*: Nilfanion

Image:SteamtrainTunnel.jpg *Source*: http://en.wikipedia.org/w/index.php?title=File:SteamtrainTunnel.jpg *License*: unknown *Contributors*: myself

Image:GWR No.7827 Lydham Manor .jpg *Source*: http://en.wikipedia.org/w/index.php?title=File:GWR_No.7827_Lydham_Manor_.jpg *License*: unknown *Contributors*: User:6024kingedward1

Image:GWR 7800 Class 7827 Lydham Manor.JPG *Source*: http://en.wikipedia.org/w/index.php?title=File:GWR_7800_Class_7827_Lydham_Manor.JPG *License*: unknown *Contributors*: User:Cavalry

Image:Diesel locomotive Paignton.jpg *Source*: http://en.wikipedia.org/w/index.php?title=File:Diesel_locomotive_Paignton.jpg *License*: unknown *Contributors*: myself

Image:Devon Belle Saloon.jpg *Source*: http://en.wikipedia.org/w/index.php?title=File:Devon_Belle_Saloon.jpg *License*: unknown *Contributors*: User:Geof Sheppard

image:churstonbldg.jpg *Source*: http://en.wikipedia.org/w/index.php?title=File:Churstonbldg.jpg *License*: unknown *Contributors*: Bjh21, Geof Sheppard, Optimist on the run

File:Churston station with auto-train for Brixham geograph-2591582-by-Ben-Brooksbank.jpg *Source*: http://en.wikipedia.org/w/index.php?title=File:Churston_station_with_auto-train_for_Brixham_geograph-2591582-by-Ben-Brooksbank.jpg *License*: unknown *Contributors*: Ben Brooksbank

Image:HR icon.svg *Source*: http://en.wikipedia.org/w/index.php?title=File:HR_icon.svg *License*: unknown *Contributors*: Bayo, Beao

GNU Free Documentation License Version 1.2, November 2002 Copyright (C) 2000,2001,2002 Free Software Foundation, Inc. 59 Temple Place, Suite 330, Boston, MA 02111-1307 USA Everyone is permitted to copy and distribute verbatim copies of this license document, but changing it is not allowed.

0. PREAMBLE
The purpose of this License is to make a manual, textbook, or other functional and useful document "free" in the sense of freedom: to assure everyone the effective freedom to copy and redistribute it, with or without modifying it, either commercially or noncommercially. Secondarily, this License preserves for the author and publisher a way to get credit for their work, while not being considered responsible for modifications made by others. This License is a kind of "copyleft", which means that derivative works of the document must themselves be free in the same sense. It complements the GNU General Public License, which is a copyleft license designed for free software. We have designed this License in order to use it for manuals for free software, because free software needs free documentation: a free program should come with manuals providing the same freedoms that the software does. But this License is not limited to software manuals; it can be used for any textual work, regardless of subject matter or whether it is published as a printed book. We recommend this License principally for works whose purpose is instruction or reference.

1. APPLICABILITY AND DEFINITIONS
This License applies to any manual or other work, in any medium, that contains a notice placed by the copyright holder saying it can be distributed under the terms of this License. Such a notice grants a world-wide, royalty-free license, unlimited in duration, to use that work under the conditions stated herein. The "Document", below, refers to any such manual or work. Any member of the public is a licensee, and is addressed as "you". You accept the license if you copy, modify or distribute the work in a way requiring permission under copyright law. A "Modified Version" of the Document means any work containing the Document or a portion of it, either copied verbatim, or with modifications and/or translated into another language. A "Secondary Section" is a named appendix or a front-matter section of the Document that deals exclusively with the relationship of the publishers or authors of the Document to the Document's overall subject (or to related matters) and contains nothing that could fall directly within that overall subject. (Thus, if the Document is in part a textbook of mathematics, a Secondary Section may not explain any mathematics.) The relationship could be a matter of historical connection with the subject or with related matters, or of legal, commercial, philosophical, ethical or political position regarding them. The "Invariant Sections" are certain Secondary Sections whose titles are designated, as being those of Invariant Sections, in the notice that says that the Document is released under this License. If a section does not fit the above definition of Secondary then it is not allowed to be designated as Invariant. The Document may contain zero Invariant Sections. If the Document does not identify any Invariant Sections then there are none. The "Cover Texts" are certain short passages of text that are listed, as Front-Cover Texts or Back-Cover Texts, in the notice that says that the Document is released under this License. A Front-Cover Text may be at most 5 words, and a Back-Cover Text may be at most 25 words. A "Transparent" copy of the Document means a machine-readable copy, represented in a format whose specification is available to the general public, that is suitable for revising the document straightforwardly with generic text editors or (for images composed of pixels) generic paint programs or (for drawings) some widely available drawing editor, and that is suitable for input to text formatters or for automatic translation to a variety of formats suitable for input to text formatters. A copy made in an otherwise Transparent file format whose markup, or absence of markup, has been arranged to thwart or discourage subsequent modification by readers is not Transparent. An image format is not Transparent if used for any substantial amount of text. A copy that is not "Transparent" is called "Opaque". Examples of suitable formats for Transparent copies include plain ASCII without markup, Texinfo input format, LaTeX input format, SGML or XML using a publicly available DTD, and standard-conforming simple HTML, PostScript or PDF designed for human modification. Examples of transparent image formats include PNG, XCF and JPG. Opaque formats include proprietary formats that can be read and edited only by proprietary word processors, SGML or XML for which the DTD and/or processing tools are not generally available, and the machine-generated HTML, PostScript or PDF produced by some word processors for output purposes only. The "Title Page" means, for a printed book, the title page itself, plus such following pages as are needed to hold, legibly, the material this License requires to appear in the title page. For works in formats which do not have any title page as such, "Title Page" means the text near the most prominent appearance of the work's title, preceding the beginning of the body of the text. A section "Entitled XYZ" means a named subunit of the Document whose title either is precisely XYZ or contains XYZ in parentheses following text that translates XYZ in another language. (Here XYZ stands for a specific section name mentioned below, such as "Acknowledgements", "Dedications", "Endorsements", or "History".) To "Preserve the Title" of such a section when you modify the Document means that it remains a section "Entitled XYZ" according to this definition. The Document may include Warranty Disclaimers next to the notice which states that this License applies to the Document. These Warranty Disclaimers are considered to be included by reference in this License, but only as regards disclaiming warranties: any other implication that these Warranty Disclaimers may have is void and has no effect on the meaning of this License.

2. VERBATIM COPYING
You may copy and distribute the Document in any medium, either commercially or noncommercially, provided that this License, the copyright notices, and the license notice saying this License applies to the Document are reproduced in all copies, and that you add no other conditions whatsoever to those of this License. You may not use technical measures to obstruct or control the reading or further copying of the copies you make or distribute. However, you may accept compensation in exchange for copies. If you distribute a large enough number of copies you must also follow the conditions in section 3. You may also lend copies, under the same conditions stated above, and you may publicly display copies.

3. COPYING IN QUANTITY
If you publish printed copies (or copies in media that commonly have printed covers) of the Document, numbering more than 100, and the Document's license notice requires Cover Texts, you must enclose the copies in covers that carry, clearly and legibly, all these Cover Texts: Front-Cover Texts on the front cover, and Back-Cover Texts on the back cover. Both covers must also clearly and legibly identify you as the publisher of these copies. The front cover must present the full title with all words of the title equally prominent and visible. You may add other material on the covers in addition. Copying with changes limited to the covers, as long as they preserve the title of the Document and satisfy these conditions, can be treated as verbatim copying in other respects. If the required texts for either cover are too voluminous to fit legibly, you should put the first ones listed (as many as fit reasonably) on the actual cover, and continue the rest onto adjacent pages. If you publish or distribute Opaque copies of the Document numbering more than 100, you must either include a machine-readable Transparent copy along with each Opaque copy, or state in or with each Opaque copy a computer-network location from which the general network-using public has access to download using public-standard network protocols a complete Transparent copy of the Document, free of added material. If you use the latter option, you must take reasonably prudent steps, when you begin distribution of Opaque copies in quantity, to ensure that this Transparent copy will remain thus accessible at the stated location until at least one year after the last time you distribute an Opaque copy (directly or through your agents or retailers) of that edition to the public. It is requested, but not required, that you contact the authors of the Document well before redistributing any large number of copies, to give them a chance to provide you with an updated version of the Document.

4. MODIFICATIONS
You may copy and distribute a Modified Version of the Document under the conditions of sections 2 and 3 above, provided that you release the Modified Version under precisely this License, with the Modified Version filling the role of the Document, thus licensing distribution and modification of the Modified Version to whoever possesses a copy of it. In addition, you must do these things in the Modified Version: A. Use in the Title Page (and on the covers, if any) a title distinct from that of the Document, and from those of previous versions (which should, if there were any, be listed in the History section of the Document). You may use the same title as a previous version if the original publisher of that version gives permission. B. List on the Title Page, as authors, one or more persons or entities responsible for authorship of the modifications in the Modified Version, together with at least five of the principal authors of the Document (all of its principal authors, if it has fewer than five), unless they release you from this requirement. C. State on the Title page the name of the publisher of the Modified Version, as the publisher. D. Preserve all the copyright notices of the Document. E. Add an appropriate copyright notice for your modifications adjacent to the other copyright notices. F. Include, immediately after the copyright notices, a license notice giving the public permission to use the Modified Version under the terms of this License, in the form shown in the Addendum below. G. Preserve in that license notice the full lists of Invariant Sections and required Cover Texts given in the Document's license notice. H. Include an unaltered copy of this License. I. Preserve the section Entitled "History", Preserve its Title, and add to it an item stating at least the title, year, new authors, and publisher of the Modified Version as given on the Title Page. If there is no section Entitled "History" in the Document, create one stating the title, year, authors, and publisher of the Document as given on its Title Page, then add an item describing the Modified Version as stated in the previous sentence. J. Preserve the network location, if any, given in the Document for public access to a Transparent copy of the Document, and likewise the network locations given in the Document for previous versions it was based on. These may be placed in the "History" section. You may omit a network location for a work that was published at least four years before the Document itself, or if the original publisher of the version it refers to gives permission. K. For any section Entitled "Acknowledgements" or "Dedications", Preserve the Title of the section, and preserve in the section all the substance and tone of each of the contributor acknowledgements and/or dedications given therein. L. Preserve all the Invariant Sections of the Document, unaltered in their text and in their titles. Section numbers or the equivalent are not considered part of the section titles. M. Delete any section Entitled "Endorsements". Such a section may not be included in the Modified Version. N. Do not retitle any existing section to be Entitled "Endorsements" or to conflict in title with any Invariant Section. O. Preserve any Warranty Disclaimers. If the Modified Version includes new front-matter sections or appendices that qualify as Secondary Sections and contain no material copied from the Document, you may at your option designate some or all of these sections as invariant. To do this, add their titles to the list of Invariant Sections in the Modified Version's license notice. These titles must be distinct from any other section titles. You may add a section Entitled "Endorsements", provided it contains nothing but endorsements of your Modified Version by various parties--for example, statements of peer review or that the text has been approved by an organization as the authoritative definition of a standard. You may add a passage of up to five words as a Front-Cover Text, and a passage of up to 25 words as a Back-Cover Text, to the end of the list of Cover Texts in the Modified Version. Only one passage of Front-Cover Text and one of Back-Cover Text may be added by (or through arrangements made by) any one entity. If the Document already includes a cover text for the same cover, previously added by you or by arrangement made by the same entity you are acting on behalf of, you may not add another; but you may replace the old one, on explicit permission from the previous publisher that added the old one. The author(s) and publisher(s) of the Document do not by this License give permission to use their names for publicity for or to assert or imply endorsement of any Modified Version.

5. COMBINING DOCUMENTS
You may combine the Document with other documents released under this License, under the terms defined in section 4 above for modified versions, provided that you include in the combination all of the Invariant Sections of all of the original documents, unmodified, and list them all as Invariant Sections of your combined work in its license notice, and that you preserve all their Warranty Disclaimers. The combined work need only contain one copy of this License, and multiple identical Invariant Sections may be replaced with a single copy. If there are multiple Invariant Sections with the same name but different contents, make the title of each such section unique by adding at the end of it, in parentheses, the name of the original author or publisher of that section if known, or else a unique number. Make the same adjustment to the section titles in the list of Invariant Sections in the license notice of the combined work. In the combination, you must combine any sections Entitled "History" in the various original documents, forming one section Entitled "History"; likewise combine any sections Entitled "Acknowledgements", and any sections Entitled "Dedications". You must delete all sections Entitled "Endorsements".

6. COLLECTIONS OF DOCUMENTS
You may make a collection consisting of the Document and other documents released under this License, and replace the individual copies of this License in the various documents with a single copy that is included in the collection, provided that you follow the rules of this License for verbatim copying of each of the documents in all other respects. You may extract a single document from such a collection, and distribute it individually under this License, provided you insert a copy of this License into the extracted document, and follow this License in all other respects regarding verbatim copying of that document.

7. AGGREGATION WITH INDEPENDENT WORKS
A compilation of the Document or its derivatives with other separate and independent documents or works, in or on a volume of a storage or distribution medium, is called an "aggregate" if the copyright resulting from the compilation is not used to limit the legal rights of the compilation's users beyond what the individual works permit. When the Document is included in an aggregate, this License does not apply to the other works in the aggregate which are not themselves derivative works of the Document. If the Cover Text requirement of section 3 is applicable to these copies of the Document, then if the Document is less than one half of the entire aggregate, the Document's Cover Texts may be placed on covers that bracket the Document within the aggregate, or the electronic equivalent of covers if the Document is in electronic form. Otherwise they must appear on printed covers that bracket the whole aggregate.

8. TRANSLATION
Translation is considered a kind of modification, so you may distribute translations of the Document under the terms of section 4. Replacing Invariant Sections with translations requires special permission from their copyright holders, but you may include translations of some or all Invariant Sections in addition to the original versions of these Invariant Sections. You may include a translation of this License, and all the license notices in the Document, and any Warranty Disclaimers, provided that you also include the original English version of this License and the original versions of those notices and disclaimers. In case of a disagreement between the translation and the original version of this License or a notice or disclaimer, the original version will prevail. If a section in the Document is Entitled "Acknowledgements", "Dedications", or "History", the requirement (section 4) to Preserve its Title (section 1) will typically require changing the actual title.

9. TERMINATION
You may not copy, modify, sublicense, or distribute the Document except as expressly provided for under this License. Any other attempt to copy, modify, sublicense or distribute the Document is void, and will automatically terminate your rights under this License. However, parties who have received copies, or rights, from you under this License will not have their licenses terminated so long as such parties remain in full compliance.

10. FUTURE REVISIONS OF THIS LICENSE
The Free Software Foundation may publish new, revised versions of the GNU Free Documentation License from time to time. Such new versions will be similar in spirit to the present version, but may differ in detail to address new problems or concerns. See http://www.gnu.org/copyleft/. Each version of the License is given a distinguishing version number. If the Document specifies that a particular numbered version of this License "or any later version" applies to it, you have the option of following the terms and conditions either of that specified version or of any later version that has been published (not as a draft) by the Free Software Foundation. If the Document does not specify a version number of this License, you may choose any version ever published (not as a draft) by the Free Software Foundation. ADDENDUM: How to use this License for your documents To use this License in a document you have written, include a copy of the License in the document and put the following copyright and license notices just after the title page: Copyright (c) YEAR YOUR NAME. Permission is granted to copy, distribute and/or modify this document under the terms of the GNU Free Documentation License, Version 1.2 or any later version published by the Free Software Foundation; with no Invariant Sections, no Front-Cover Texts, and no Back-Cover Texts. A copy of the license is included in the section entitled "GNU Free Documentation License". If you have Invariant Sections, Front-Cover Texts and Back-Cover Texts, replace the "with...Texts." line with this: with the Invariant Sections being LIST THEIR TITLES, with the Front-Cover Texts being LIST, and with the Back-Cover Texts being LIST. If you have Invariant Sections without Cover Texts, or some other combination of the three, merge those two alternatives to suit the situation. If your document contains nontrivial examples of program code, we recommend releasing these examples in parallel under your choice of free software license, such as the GNU General Public License, to permit their use in free software.

Printed by Books on Demand GmbH, Norderstedt / Germany